CLIMATE CHANGE, ETHICS AND HUMAN SECURITY

This book presents human security perspectives on climate change, raising issues of equity, ethics and environmental justice, as well as our capacity to respond to what is increasingly considered to be the greatest societal challenge for humankind. The authors – a group of international experts – represent a variety of fields and perspectives. Taken together, the chapters make the argument that climate change must be viewed as an issue of human security, rather than simply as an environmental problem that can be managed in isolation from larger questions concerning development trajectories, poverty and equity issues, human rights, and ethical and moral obligations towards the poor and vulnerable, and to future generations. The book shows that the concept of human security offers a new way of understanding the challenges of climate change, as well as the responses that can lead to a more equitable, resilient and sustainable future.

Climate Change, Ethics and Human Security will be of interest to researchers, policy makers, and practitioners concerned with the human dimensions of climate change, as well as to upper-level students in the social sciences and humanities who are interested in climate change.

Dr Karen L. O'Brien is a Professor in the Department of Sociology and Human Geography at the University of Oslo, Norway, and Chair of the Global Environmental Change and Human Security (GECHS) project of the International Human Dimensions Programme on Global Environmental Change (IHDP). Her research focuses on climate change vulnerability and adaptation, on interactions between globalisation and climate change, and on the role that values and world views play in responding to environmental change. She currently leads a large social science-based project on the Potentials of and Limits to Climate Change Adaptation in Norway (PLAN). She has recently published a book with Robin Leichenko on *Environmental Change and Globalization: Double Exposures* (Oxford University Press, 2008), which received the 2008 AAG Meridian Book Award for the Outstanding Scholarly Work in Geography. She was a lead author on the Intergovernmental Panel on Climate Change (IPCC) *Fourth Assessment Report* and the *Special Report on Managing the Risks of Extreme Events and Disasters to Advance Climate Change Adaptation*. She is on the editorial board of *Global*

Environmental Change, the *Annals of the Association of American Geography* and *Ecology and Society*, and on the steering committee of the Comparative Research Programme on Poverty (CROP), the Norwegian Global Change Committee and Concerned Scientists – Norway.

Asunción Lera St.Clair is Professor of Sociology at the University of Bergen, Norway, and Scientific Director of the Comparative Research Programme on Poverty (CROP), one of the leading programmes of the International Social Science Council (ISSC); co-leader (with Victoria Lawson) of the WUN Critical Global Poverty Studies Research group; Vice-President of the International Development Ethics Association (IDEA); board member of the Rafto Foundation for Human Rights and a member of the editorial boards of various international journals and other international organisations. Her work is primarily interdisciplinary, focusing on ethical issues related to development and poverty with a special focus on human rights and attempting to provide alternative perspectives on global poverty. St.Clair has also focused attention on the challenges posed by climate change in relation to poverty and development studies, aiming to offer holistic ethically grounded perspectives on both challenges. She has recently published in the journals of *Global Governance*, *Global Social Policy*, *Globalisations* and *Global Ethics*. Her most recent publications are *Global Poverty, Ethics, and Human Rights: The Role of Multilateral Institutions* (2009, Routledge, co-authored with Desmond McNeill) and *Development Ethics: A Reader* (2010, Ashgate, co-edited with Des Gasper).

Berit Kristoffersen is a political geographer and PhD student at the Department of Political Science, University of Tromsø, Norway. Her PhD research is on human and environmental security in Norway, in the context of the state and industry's strategies for petroleum development in Arctic territories. She is a research fellow in the research programme, The Potentials of and Limits to Climate Change Adaptation in Norway (PLAN), where she is also working on a project on values and climate change in Norway. Before starting a PhD she was involved in Norwegian organisations and social movements emerging in the post-Seattle globalisation movement and worked as a freelance writer and editor of several reports and book chapters for NGOs, such as Forum for Environment and Development, Zero, Adbusters Norway, and Attac.

CLIMATE CHANGE, ETHICS AND HUMAN SECURITY

Edited by

KAREN O'BRIEN
University of Oslo, Norway

ASUNCIÓN LERA ST.CLAIR
University of Bergen, Norway

BERIT KRISTOFFERSEN
University of Tromsø, Norway

CAMBRIDGE
UNIVERSITY PRESS

University Printing House, Cambridge CB2 8BS, United Kingdom

Published in the United States of America by Cambridge University Press, New York

Cambridge University Press is part of the University of Cambridge.

It furthers the University's mission by disseminating knowledge in the pursuit of education, learning and research at the highest international levels of excellence.

www.cambridge.org
Information on this title: www.cambridge.org/9781107695856

First published 2010
First paperback edition 2014

A catalogue record for this publication is available from the British Library

Library of Congress Cataloguing in Publication data
Climate change, ethics and human security / [edited by] Karen O'Brien, Asuncion Lera St. Clair,
Berit Kristoffersen.
p. cm.
ISBN 978-0-521-19766-3 (hardback)
1. Human beings – Effect of climate on. 2. Human security. 3. Human rights. 4. Climatic changes – Social
aspects. I. O'Brien, Karen L. II. St. Clair, Asuncion Lera. III. Kristoffersen, Berit. IV. Title.
GF71.C56 2010
304.2'3–dc22
2010015149

ISBN 978-0-521-19766-3 Hardback
ISBN 978-1-107-69585-6 Paperback

Contents

Contributors

W. Neil Adger Professor in the Tyndall Centre for Climate Change Research, School of Environmental Sciences, University of East Anglia, Norwich, UK.

Jon Barnett Associate Professor at the Australian Research Council and Fellow and Reader, Department of Resource Management and Geography, University of Melbourne, Australia.

Livia Bizikova Project Manager at the International Institute for Sustainable Development, Winnipeg, Canada.

Sarah Burch Visiting Research Associate, Environmental Change Institute, University of Oxford, UK.

Simon Caney Professor of Political Theory and Fellow and Tutor in Politics at Magdalen College, University of Oxford, UK.

Stewart Cohen Senior Researcher with the Adaptation and Impacts Research Division (AIRD), Environment Canada, and Adjunct Professor with the Department of Forest Resources Management, University of British Columbia (UBC), Canada.

Stephen M. Gardiner Associate Professor at the Department of Philosophy, University of Washington, Seattle, USA.

Des Gasper Professor at the International Institute of Social Studies, The Hague, a university institute within Erasmus University Rotterdam, The Netherlands.

Heide Hackmann Secretary-General of the International Social Science Council (ISSC), Paris, France and Associate Research Fellow, Centre for Social Science Research, University of Cape Town, South Africa.

Bronwyn Hayward Political Scientist at the University of Canterbury, New Zealand and Visiting Fellow at the Tyndall Centre for Climate Change Research and the RESOLVE Centre for the Study of Environmental Values, Attitudes and Lifestyle Change at the University of Surrey, UK.

Berit Kristoffersen Research Fellow, Department of Political Science, University of Tromsø, Norway.

Desmond McNeill Research Professor at Centre for Development and the Environment (SUM), University of Oslo, Norway.

Donald R. Nelson Assistant Professor at the Department of Anthropology, University of Georgia, and Visiting Fellow, Tyndall Centre for Climate Change Research University of East Anglia, UK.

Karen O'Brien Professor in the Department of Sociology and Human Geography, University of Oslo, Norway, and Chair of the Global Environmental Change and Human Security Project of the IHDP.

Bradley C. Parks PhD Candidate in the Department of International Relations at the London School of Economics and Political Science and Research Associate at the College of William and Mary's Institute for Theory and Practice of International Relations Virginia, USA.

John Robinson Professor at Institute of Resources, Environment and Sustainability, University of British Columbia, Vancouver, Canada.

J. Timmons Roberts Professor of Sociology and Environmental Studies and Director of the Center for Environmental Studies at Brown University in Rhode Island, USA.

Asunción Lera St.Clair Professor, Department of Sociology, and Scientific Director, Comparative Research Programme on Poverty (CROP), University of Bergen, Norway.

Foreword

In May 2009 the International Social Science Council (ISSC) convened the first ever World Social Science Forum.[1] The theme of the Forum – One Planet, Worlds Apart? – challenged social scientists from different parts of the world, working with different theories and different methodologies, to join forces in tackling the most important global problems of the day, and to do so in ways that make sense of shifting geopolitics, address global inequalities and preserve human culture, dignity and diversity.

Can science save us from climate change? This was one of the key questions posed at the Forum. Those asked to address it included the Nobel laureate Rajendra Pachauri, Chair of the Intergovernmental Panel on Climate Change (IPCC) and Roberta Balstad, Co-Director of the Center for Research on Environmental Decisions at Columbia University and Editor-in-Chief of *Weather, Climate, and Society*, a new journal of the American Meteorological Society. Both speakers issued a clear and concise message: climate change research needs a stronger social science voice; more than that, to produce the kind of knowledge we need to respond effectively to the complexities of global environmental change, an integration of natural and social sciences is no longer a choice but a simple necessity.

ISSC President Gudmund Hernes reminds us that today we know that climate change is not about 'the forces of nature, so to speak, autonomously at work, like planetary motions'; we know that what has set those forces in motion is human action. The key causes of climate change are primarily social and the grave consequences of such change will also be social. 'Land for agriculture will be destroyed by inundations and drought. Poverty will increase. Water and food will be in shorter supply. Diseases will spread. Social inequality will be sharpened. Migration will mount from climate change refugees. Social crises can multiply, and

---

[1] The Forum was held in Bergen, Norway, from 10 to 12 May 2009. It was hosted by the University of Bergen and co-organised by the University's Stein Rokkan Centre for Social Studies.

conflicts may be provoked.'[2] In these circumstances we can no longer afford to talk only about natural phenomena but must talk also – and urgently – about human behaviour, about human perceptions, values and rights, human responses and responsibilities.

Climate Change, Ethics and Human Security talks exactly about these things. In doing so it places human beings – individuals and communities – at the centre of analysis and eliminates once and for all remaining doubts that social – and human – scientific knowledge is necessary knowledge for the future of our planet. The book draws attention to a wide range of new important questions that the social sciences and humanities bring to the climate change research agenda. And it insists on an integral approach to tackling such questions; an approach that considers both subjective and objective dimensions of climate change and incorporates both qualitative and quantitative methodologies. The value of this approach is clear in the way that it serves to place poverty and the poor at the centre of our understanding not only of the risks posed by climate change but also of our responses to it. This in turn necessitates a fundamental reassessment of standard development models and cautions against uncritically accepting those ideas about poverty that perpetuate them. This kind of connected thinking, the intellectual approach that facilitates it and the new frames of reference that it provides, creates much needed space for innovative, alternative knowledge on major issues, like the links between climate change and poverty. Such knowledge must be incorporated in international assessments like the IPCC Fifth Assessment Report.

Turning more specifically to the integration of the type of social science and humanities knowledge forwarded here with that produced by the natural sciences, it is not always entirely self-evident what this means and how best it can be accomplished. The ISSC is taking the call for integrated research seriously and has committed itself to working across lines of division between the sciences to the benefit of our common humanity and shared physical environment. In this, the International Council for Science (ICSU) and, increasingly, the International Council for Philosophy and Humanistic Studies (CIPSH), are key strategic partners.

ICSU and the ISSC share a positive history of collaboration in the field of global environmental change research. Recognising that a polarisation between social and natural sciences serves only as an obstacle to addressing key global problems, the two organisations agreed in 1996 to co-sponsor the International Human Dimensions of Global Environmental Change Programme (IHDP).[3] Without diminishing the value of the IHDP experience to date, both organisations now recognise

---

[2] Hernes, G. (2009). One planet – two cultures? *Public Service Review: Science and Technology*, 2, 54–5.
[3] The ISSC and ICSU established the IHDP in 1996; it originated from what had been called the Human Dimensions Programme (HDP), which was launched by the ISSC in 1990. Since 2007, the IHDP has been Co-sponsored by the United Nations University.

the need to take collaboration between the sciences to a deeper, more constructive and complementary form. This means moving beyond multi- or even interdisciplinary collaboration. It certainly means moving beyond the idea that some sciences, or some disciplines, should serve others; that they should wait in the margins to assist with the translation and take-up of research findings. It means promoting integrated research: research that in its very design, execution and application demands the joint efforts of natural and social scientists.

This book makes the strongest possible case for integrated research on climate change, also drawing in the humanities. More than that, it brings to the integration imperative two essential insights; lessons that will equip us to make integration a reality in the promotion, funding, practice and evaluation of climate change research. The first lesson is that integration demands an openness to asking new questions, different questions, invisible questions. Integration does not, in other words, mean getting social scientists to join in attempts at addressing problems, which have largely, if not solely, been framed by natural scientists. Framing climate change as an issue of human security does not negate the importance of those problems. If anything, it enhances our understanding of them. It also allows us to better inform the likely consequences of the policy choices made to address those problems. And, perhaps most importantly of all, it urges us to recognise that in addition to the fact that the causes and consequences of climate change are primarily social, so must the solutions be.

The second lesson concerns the fragmentation of the social sciences themselves. When it comes to climate change there seems to be not one social science but many. Again, the issue of framing is critical. What type of social science sets the climate change research agenda? Whose research questions are being asked? What theoretical approaches and methodologies dominate debates? By raising these questions, *Climate Change, Ethics and Human Security* raises fundamental questions – not least of all for the ISSC – about the need to define common tasks and set shared agendas within the social sciences. With our planet imperilled, with deep inequalities evident within and across countries, with vulnerability to poverty increasing and with persistent severe poverty a reality, can the social sciences afford to work as worlds apart? And with expectations that science can indeed save us from climate change higher than ever before, can we afford, as social scientists, not to speak with one voice?

The effects of climate change are inescapable and relentless. They pose severe challenges to all human beings from all parts of the world. If science is to play a role in meeting these challenges, scientists have to get their act together. And they have to do so jointly, across disciplinary and organisational boundaries, across issues and methodologies, across national and regional borders. This book shows us what that means, and points the way forward towards the kind of integration of knowledge that climate change demands of us.

For the ISSC it is particularly important – and gratifying – to see knowledge networks from two of its primary international programmes – the Comparative Research Programme on Poverty (CROP), on one hand, and the International Human Dimensions of Global Environmental Change Programme (IHDP), on the other hand – coming together to tackle the single most important issue facing our planet today. This book, which results from collaborative work between the leaders of CROP and one of the IHDP's core projects – the Global Environmental Change and Human Security (GECHS) project – is an example of the innovations that come from joint efforts.

Heide Hackmann
Secretary-General
International Social Science Council (ISSC)
Paris, France

June 2009

Preface

This book is the result of a European Science Foundation (ESF) Exploratory Workshop on 'Shifting the Discourse: Climate Change as an Issue of Human Security', which was held 21–23 June 2007 in Oslo, Norway. This was the same year that the IPCC Fourth Assessment Report was published, and the year that both the IPCC and former US Vice President Al Gore received the Nobel Peace Prize for their work on climate change. Since then, the amount of attention paid to climate change has increased dramatically, particularly in the run-up to the 2009 Conference of the Parties (COP) in Copenhagen, where new international agreements to address climate change will be discussed. Yet, although the connection to peace and security was recognised in 2007 and many more voices and perspectives can now be heard, the discourse itself on climate change has not changed significantly. It is still framed as an environmental problem that can be managed through international agreements for emissions reductions, through market mechanisms for carbon management and through technological advances that will create clean and green societies. Many voices, including some that have long been sceptical about climate change, are now advocating geo-engineering as a solution. The institutionalised, mainstream discourse on climate change has not recognised it as an issue that is first and foremost about the security of individuals and communities and their relationship with the world around them, which includes responsibilities to one another, to other species and to future generations. As the contributions to this book make clear, the equity, justice and ethical dimensions of the problem must be included, and voices from the social sciences, humanities and other fields must be heard. We argue that a more integral understanding of the problem and solutions associated with climate change can only be forged by developing a 'new science' on climate change – a science that recognises that the drivers and consequences of climate change go far beyond what can be measured by econometrics and statistics alone. This new science also recognises the normative dimensions of climate change and the non-material aspects that are differentially valued, yet play an important role in culture and

xiii

human experiences. As we approach 2010, the time has come to recognise that climate change is as much an ethical issue as it is a 'scientific' issue, and it cannot be decoupled from debates about the ethical demands posed by equitable development, the feasibility of the eradication of poverty, sustainability and the way that we as human beings perceive of and create the future.

We are grateful to many individuals and institutions for their support of this book project. In particular, we thank the sponsors of the 2007 workshop, which, in addition to ESF, include the Ethics Programme of the University of Oslo and the Global Environmental Change and Human Security (GECHS) project, one of the core projects of the International Human Dimensions Programme (IHDP). GECHS is based at the University of Oslo and is funded by the Norwegian Research Council, with generous additional support from Norad and the Norwegian Royal Ministry of Foreign Affairs. We would also like to thank all of the workshop participants for their comments and discussions, even if not all could contribute to this book. We would specifically like to thank the two keynote speakers, Henry Shue from Oxford University and Helge Drange from the University of Bergen and the Bjerknes Centre for Climate Research.

We thank the staff of the GECHS International Project Office, especially Lynn Rosentrater and Kirsten Ulsrud for their assistance in organising the workshop, and Linda Sygna and Øystein Kristiansen for post-workshop support and editorial assistance. Many thanks go to Kristian Stokke for his assistance with the tables and figures. We also thank four anonymous reviewers for very insightful comments and suggestions based on the original book proposal. We are grateful to Matt Lloyd, Christopher Hudson and Laura Clark at Cambridge University Press for support, professional advice and, most of all, continued patience. Finally, we thank our friends and families, and would like to dedicate this book to our children, Jens Erik, Espen, Annika, Thomas, Nicholas and future generations.

Karen O'Brien,
Asunción Lera St.Clair and
Berit Kristoffersen
November 2009
Bergen and Oslo, Norway

Part I

Framings

1

The framing of climate change: why it matters

KAREN O'BRIEN, ASUNCIÓN LERA ST.CLAIR AND
BERIT KRISTOFFERSEN

Introduction

Climate change is now considered by many to be the most complex and serious environmental issue that human societies have ever faced. The science is unequivocal – human activities are influencing the climate system, contributing to increases in global average air and ocean temperatures, the widespread melting of snow and ice, and rising average global sea levels (IPCC, 2007). Well-known economists have shown that there are instrumental reasons to immediately minimise CO_2 emissions (Stern, 2007), and these arguments are underscored by global assessments of the potential human impacts (UNDP, 2007/2008; Global Humanitarian Forum, 2009). Some voices argue that climate change is a cultural phenomenon that is reshaping understandings of humanity's place on Earth (Hulme, 2009), while others warn that '[w]e do not seem to have the slightest understanding of the seriousness of our plight' (Lovelock, 2009: 4). Al Gore's famous statement that 'the truth about the climate crisis is an inconvenient one that means we are going to have to change the way we live our lives' (2006: 286) captures the essence of the climate change challenge. The problem is that we have very little idea about what exactly needs to be changed and why.

Although it has taken the global community and the general public many years to acknowledge the inconvenient truth pointed to by Gore, the urgency of responding to the climate crisis is becoming increasingly evident. This has led to a wide range of proposed responses, ranging from a 'one degree war' plan, to strategies to 'overshoot, adapt and recover' (Cambridge Programme for Sustainability Leadership, 2009; Parry et al., 2009). Yet, perhaps it is the time to acknowledge that Gore's inconvenient truth is not the whole truth. A set of uncomfortable truths that have not yet been widely acknowledged is related to questions that have not yet been widely asked or answered. These questions include: What types of meaningful changes and alternative futures should be envisioned and why? How is this process going to happen? What types of risks are bearable, and by whom? How can transformations

Climate Change, Ethics and Human Security, eds. Karen O'Brien, Asunción Lera St.Clair and Berit Kristoffersen.
Published by Cambridge University Press. © Cambridge University Press 2010.

be managed to minimise injustices and conflicts? And most importantly, who is the 'we' that is really going to have to change?

The answers to these questions, which are only starting to emerge, make it clear that climate change is not simply an environmental issue that can be managed through behavioural changes, sectoral interventions or new regulations. It is not a problem that can be addressed single-handedly by environmental ministries, by international institutions and non-governmental organisations or by development aid and adaptation funds. Finally, it is not a problem that can be *solved* by ecological modernisation, ecosystem stewardship or sustainable development. It is, instead, a problem that can only be *resolved* by focusing on climate change as an issue of human security, which includes a thorough investigation of what it means for humans to be 'secure'. This demands, first and foremost, a change in the way that we think about change. It requires a shift away from the dominant framing that focuses on responding to change through a utilitarian, problem-solving approach or cost–benefit analyses, and towards a framing that recognises and prioritises the capacity of individuals and communities to both respond to and create change, including envisioning and pursuing alternative futures.

In this book, we explore some less familiar, yet important questions related to climate change, including issues of framings, equity, ethics and reflexivity. Questioning the framing of climate change matters. It matters because dominant perspectives do not confront fundamental aspects of the problem and may lead to regretful (and deadly) actions or inaction. Indeed, the current discursive orientation on climate change focuses disproportionately on regulations, policies and behavioural changes, which alone are unlikely to address or influence the underlying factors that threaten the capacity of individuals and communities to respond to threats to their social, human and environmental rights. Many of the so-called 'solutions' to climate change are partial responses to the symptoms; they fail to address the underlying and structural conditions necessary for resolving the problem, i.e. for creating transformational change. As the contributions to this volume show, shifting the dominant framing of climate change towards a focus on human security raises questions of ethics, values, justice and responsibility.

In the last decade, human security has emerged as both a concept and a discourse that complements the closely related notions of human development and human rights (Gasper, 2005; and Chapter 2). Human security has been defined in very general terms as freedom from want and freedom from fear, and more specifically as having the ability to respond to critical and pervasive threats (UNDP, 1994; GECHS, 1999; Commission on Human Security, 2003). It is a concept that is centred on people and their social relations, rather than on national and state security needs (see Barnett, 2001b; Dalby, 2002, 2009). Human security addresses the wellbeing of individuals from multiple and interrelated perspectives: income security, food security, health security, environmental security, community/identity security and security of political freedoms. It is inherently an integrative and relational concept that draws attention to present and emerging vulnerability that is generated through dynamic social, political, economic, institutional, cultural and technological conditions and their historical legacies.

The discourse on human security invokes normative claims that 'what matters is the content of individuals' lives, including a reasonable degree of stability' (Gasper, 2005: 228). The concept of human security, broadly understood and closely inter-related to norms, values, rights and entitlements, draws attention to notions of empowerment, protection and responsibilities. In other words, human security is about the protection and fulfilment of people's vital freedoms and the development of capabilities to create satisfying lives for all people (Sen, 1999; Commission on Human Security, 2003). It also directs attention to the role of values, beliefs and world views, which are fundamental to both understanding and addressing threats and opportunities linked to climate change (Chapter 12). It takes as a point of departure the intrinsic value of the dignity of all human beings in a holistic way that includes their dependency and their relations with the natural environment, and it holds that the basic needs of any individual are neither to be sacrificed nor discounted (see Caney, Chapter 7). At the same time, it is a broad concept that embraces ideas that most cultures can relate to, albeit through different interpret-ations. Although diverse actors and users may interpret the concept of human security in different ways, it nonetheless permits a joint understanding and guidance for action (St.Clair, 2006b). As a normative discourse, human security offers a basis for fair decision making (Adger and Nelson, Chapter 5). It raises issues that are often swept aside in international scientific and policy debates about climate change, and forces a rethinking of political systems and even political theory, which may be outdated and unable to respond, and thus in need of reformulation (see Gardiner, Chapter 8; Hayward and O'Brien, Chapter 11).

In this introductory chapter, we first consider how the framing of an issue defines the scope for debates and actions. We then discuss the dominant framing of climate change, which is based on the conceptualisation of humans as related to (or coupled to) the environment, yet nonetheless separate and distinct. We consider the limits to this approach, particularly how this framing excludes key perspectives related to equity, ethics and reflexivity. The 'environmental' discourse pays little attention to people's positionality, or their values, beliefs and world views, and it ignores the importance of equity and global solidarity in both adaptation and mitigation. We next present the concept of human security as an alternative way of framing the challenges of climate change. Returning to the themes of equity, ethics and reflexivity, we consider what the emerging normative discourse on human security brings to research, debates and policy. We then present an overview of the key arguments made by the different contributors to this volume which, taken together, can be seen as an important first step towards building an alternative framework and a new science of climate change.

Framings of climate change: what are the boundaries?

Framing is a variation of discourse analysis, but also a way to situate knowledge and to interpret and question processes of knowledge formation (Jasanoff and Wynne, 1998; Forsyth, 2003; Jasanoff and Martello, 2004). Taking as a point of departure the question of how a particular issue is framed can unveil, even if only partially, some of the underlying premises, assumptions and baggage carried in all processes of knowledge production. Framing situates these processes as parts of ongoing social relations, and thus sheds light on the ways in which power relations translate into dominant expert views.

The way that a particular issue is framed is of utmost importance because it provides concrete suggestions for action, and serves as a guide for policy making (Forsyth, 2003). All climate change knowledge is increasingly (and dangerously) driven to 'hurried' and highly compromised and politicised policy decisions. As Miller and Edwards (2001: 3–4) rightly argue, contemporary debates about climate science are 'in the long run just as importantly helping to set basic rules of standing and legislation for global environmental decision making.' In the same way that expert knowledge about global poverty co-produces both knowledge and politics, climate change knowledge co-produces a particular politics of poverty and vulnerability reduction (St.Clair, 2006a, 2006b). The relevance of framing is fundamental with a problem such as climate change, which is being addressed through multiple and interacting scales of governance (Young *et al.*, 2008). A focus on framing permits the identification of disconnects, incongruities and competing views on the issue from different perspectives (e.g. local versus global). As Martello and Jasanoff (2004: 22) note, '[w]hen national-level actors confront transnational problems such as climate change, they often discover incongruities between globally constructed framings of environmental phenomena and their own histories, political cultures, and priorities.'

Highly politicised, complex and ill-structured global problems, plagued with uncertain outcomes, are precisely those that call for a careful questioning of framings. In this volume, we use the notion of framing to challenge the views offered by what can be considered the dominant 'environmental' discourse, unveiling some of its limitations, particularly in the ways that it drives actions and policy making in specific directions, bypassing alternative pathways. We consider how framing influences the ways that climate change is understood, interrogated and narrated, and how it inhibits the asking of uncomfortable questions that may nonetheless be necessary to ask, in order to understand why climate change really matters.

Climate change as a separate box

Climate change is considered to be a serious *environmental* problem. Environment, in this sense, is defined as 'the complex of physical, chemical, and biotic factors (as climate, soil, and living things) that act upon an organism or an ecological community and ultimately determine its form and survival.'[1] Within the global change research community, the atmosphere, oceans, ice, land, water, vegetation and species of all types are considered to be key components of the global environment, or Earth System. Humans are also considered to be an important part of this system, as they both drive and are impacted by environmental change (Steffen *et al.*, 2004). Nonetheless, humans are conceptualised as separate from the environment, leading to what is referred to as 'society–nature dualism' (Castree, 2005). This dichotomised or dualistic understanding and interpretation of the relationship between humans and nature underpins many debates about the causes and consequences of climate change (see Castree, 2005; Hulme, 2009). On the one hand, this dualistic understanding of nature–society relationships gives rise to a climate system that is separate and external to human activities, which may help to explain why some people deny that climate change is a problem, or attribute observed change to natural or supernatural forces outside of human control. On the other hand, this understanding may promote a strong sense of control, and a view that human influences on the climate system can be managed through the right regulations and interventions. As Adger *et al.* (2006) argue, this managerial discourse dominates the climate change debate, pointing to institutional and policy failure as the ultimate cause of the problem, and technocratic interventions as the solution. Such orthodox approaches to environmental problems, Forsyth (2003) argues, fail to acknowledge the institutional basis, including language and culture, through which environmental problems are experienced.

More recently, a new scientific paradigm has tried to capture the non-dual aspects of nature–society relationships. Research on coupled social–ecological systems recognises that humans and nature are interconnected and interdependent, interacting in complex, non-linear systems (Gunderson and Holling, 2002). Much of this thinking has come from the work on ecological resilience, which challenges the 'stable equilibrium' view of ecology and considers non-linear dynamics, thresholds, uncertainty and surprise, as well as the interplay of periods of gradual and rapid change and its dynamics over different temporal and spatial scales (Folke, 2006). From this perspective, climate change represents one more factor demonstrating how human activities are altering ecosystems and ecosystem services, which in turn have implications for human wellbeing (Millennium Ecosystem Assessment, 2005).

[1] Merriam-Webster Dictionary: http://www.merriam-webster.com/dictionary/environment

While the metaphor of coupled social–ecological systems attempts to dissolve the dichotomy, it nonetheless retains the image of society and ecology as separate but interacting systems. As Castree (2005: 224) states, 'the society–nature dualism blinds us to the need for a new vocabulary to describe the world we inhabit.' The vocabulary to describe this world may readily be found in indigenous cultures and Eastern philosophies, or in deep ecology, ecosophy and more holistic world views (see, for example, Devall and Sessions, 1985; Naess and Rothenberg, 1993; Harding, 2006; Berkes, 2008; Esbjörn-Hargens and Zimmerman, 2009). These perspectives, however, are invisible within the dominant discourse on climate change.

Although the rational scientific knowledge that underlies the dominant framing of climate change has offered important insights on the impacts of climate change and has demonstrated that the changes facing society are anything but trivial, it does little to explain how individuals and communities can best respond to threats to their environmental, social and human rights, and what climate change means for human security. In fact, the environmental discourse in many ways excludes much more than it explains when it comes to understanding the human dimensions of climate change. Below, we consider how it hides important questions related to the three themes discussed in this book: equity, ethics and reflexivity.

Equity

An environmental framing of climate change has promoted a limited understanding of the equity dimensions of climate change. To the extent that it does draw attention to these issues, it is mostly in terms of a North–South divide, particularly in relation to climate change mitigation, development and sustainability. Considerable attention has been given to the uneven relationship between those responsible for emitting greenhouse gases into the atmosphere, and those who are most likely to be affected by it (Müller, 2002; Roberts and Parks, 2006). For example, the United States emits a disproportionately large proportion of carbon dioxide, in comparison to small islands in the Pacific, which are likely to disappear if sea level rises in the next centuries (Barnett, 2001a). Less attention has been paid to equity issues within national boundaries, or those that manifest at diverse scales and units of analysis (O'Brien and Leichenko, 2006). These include many of the inequities related to race, gender, caste, ethnicity and class. The inequities that are associated with climate change are closely linked to existing inequities, and they cannot be divorced from the very processes that create these in the first place. The equity dimensions of climate change are not limited to questions of historical responsibility for greenhouse gas emissions, but encompass a much broader range of questions about the underlying and often inequitable factors that contribute to vulnerability.

Current North–South relations treat underdevelopment and poverty as issues separated from the histories of development that advanced economies have pursued or their current prioritised development paths. These paths have reinforced inequality both within and among many countries (UN, 2005). Climate change responses have tended to follow this pattern. Over the past years, there has been a rapid reorganisation of development aid bureaucracies as they have sought to include and mainstream climate change into their technical and expert work in relation to the global South (Klein *et al.*, 2007). Adaptation funds are being sought to supplement existing development funds, marking new efforts to 'climate-proof' development. Most of these efforts are reinventing or reinforcing decades-old development and poverty reduction strategies that have framed these issues as managerial matters, mainly dealt with and defined by outside experts, and driven by technocratic and economist perspectives. Poverty and development have been framed outside social relations, ignorant of the real problems of poor people and their positionality (Lawson and St.Clair, 2009).

Although much progress has been made in terms of learning how to enable and promote good development, aid continues to be driven by charitable, moralistic and top-down expert knowledge that frames poverty as separate from power and social relations, or as the geographically self-contained problems of poor countries. Dominant framings have constructed the issues as problems with an economic fix, while, at the same time, there has not been a substantial commitment on the side of wealthy countries to invest the needed funds (St.Clair, 2006a, 2006b, 2006c; McNeill and St.Clair, 2009). Not surprisingly, global commitments such as the Millennium Development Goals are both insufficient and unlikely to be met, and eliminating severe poverty remains one of the biggest moral challenges of our time (Pogge, 2004).

Ethics

As in debates about poverty reduction, many very important ethical aspects of climate change are treated as externalities in contemporary debates about environmental change. This is particularly true in relation to ethical questions about justice and fairness in climate change (see Adger *et al.*, 2006). It is not possible to quantitatively 'measure' the ethical impacts that climate change is posing for vulnerable individuals and groups. Nor is it possible to establish a fair price on the 'value' of future generations. Such calculations are perversions of a particular type of expert knowledge, emphasised and driven by political actors who have difficulty coping with the complexity of the issue in relation to the short-term demands of public service and the desire for concrete fixes and measurable results.

The challenges of climate change pose important ethical and moral questions to a global community that has substantial scientific knowledge about the trends and consequences of climate change, including projections of ecosystem changes, increased morbidity and death, massive displacements due to sea-level rise, and other impacts (see Parry *et al.*, 2007). Yet many members of this global community refuse, resist or prevent politically challenging decisions and actions from being taken to avoid dangerous climate change.

Current attitudes of scientists regarding possibilities for averting such change are pessimistic: increasingly, society is being told to prepare to adapt to temperature changes of 4°C or more over the next 100 years (Parry *et al.*, 2009). Visions of a planet in crisis, unable to sustain more than one billion people by the end of this century, are becoming common features in the news (see Lovelock, 2009). Notably absent is a vision of a more just and sustainable world built on an economy that is not based on carbon, where values associated with universalism and benevolence are prioritised. The ethical implications of these two contrasting visions are enormous, and key questions that are excluded by an environmental framing include: 'Whose vision is being pursued by society and why?' and more importantly, 'Whose values count?'

Reflexivity

It is worth questioning whether the framing of climate change as an environmental issue can, in fact, lead to the changes necessary to avoid dangerous climate change. Interpreting climate change as an environmental issue, where the 'environment' is separate from humans, prevents the self-reflection necessary to initiate large-scale transformations. In development psychology, changes in perspectives or consciousness arise when 'subject' becomes 'object' – in other words, when it becomes possible to look objectively at an event or process and reflect on it from a new and broader perspective, without being enmeshed in subjectivity (Kegan, 1994).

While scientific rationalism has mastered the objective study of the environment, many scientists themselves remain trapped in their own subjectivity, constrained by a modern world view that sees humans as separate from the environment. This limits reflexivity to the objective world of the Earth System, which includes objective analyses of the impacts of human activities on that system, as well as objective assessments of the consequences for human society. The perspective of coupled social–ecological systems likewise enables an objective analysis on the systemic interactions between society and nature, but nonetheless seldom includes an analysis of subjective and intersubjective dimensions of these systems. Neither perspectives allow for objective reflexivity on a 'bigger picture' based on non-dual human–environment relationships. Post-modern world views, as well as Beck's 'reflexive modernisation', allow for reflections on the human–environment dichotomy, including critiques on the social construction of nature, and of the wider systems and power arrangements that create problematic nature–society relationships in the first place (Beck *et al.*, 1994; Castree, 2005).

The dominant framing of climate change holds it in a separate box, as an issue that can be addressed through environmental policies and by changing individual behaviours (Maniates, 2002). This 'box' hides the diversity of motivations and interests that favour keeping things as they are, allowing for and even promoting changes in the environment in order to gain or maintain power, dominance, economic growth, familiar consumption patterns and so on. It also hides implicit assumptions and interests behind some of the rapidly emerging adaptation policies that are being propounded by governments and institutions at all scales. These implicit assumptions were perhaps best articulated by Paolo Freire in *Pedagogy of the Oppressed* (1970: 76, emphasis added):

The educated individual is the *adapted* person, because she or he is better 'fit' for the world. Translated into practice, this concept is well suited to the purposes of the oppressors, whose tranquillity rests on how well people fit the world the oppressors have created, and how little they question it. The more completely the majority *adapt* to the purposes which the dominant minority prescribe for them (thereby depriving them of the right to their own purposes), the more easily the minority can continue to prescribe.

In other words, the more completely people adapt to climate change, the more easily humans can continue to change the climate. If people do not identify what climate change means for the things that they value, reflect on how it influences or interacts with their beliefs and world views, and critically question and contest the drivers of climate change itself, then dangerous climate change is likely to be accepted as a given. With climate change accepted as a given, the solution will likely be quick fixes that permit development to proceed as usual, along with increased rates of greenhouse gas emissions.

History shows that narrow and unreflexive approaches have failed to resolve other issues, including the abolition of severe poverty. The framing of climate change as an environmental problem prevents the capacity to reflect on centuries long assumptions about development, progress and the good life. As has been the case for decades with the management of poverty and underdevelopment, managing climate change is becoming a technocratic issue; a question that can be solved with the appropriate 'fix'. In short, this dominant framing not only prevents ethical reflection, it displaces responsibility.

Opening the box: climate change as an issue of human security

There is no denying that climate change is a serious environmental issue. Yet treating the environment as a reified, independent category without questioning the ways that past and present social processes and power influence people's livelihoods and life chances can lead to inappropriate understandings of the problems likely to arise due to climate change. Of course, 'environment' can easily be defined as more than just 'nature'. It can also be considered as 'the aggregate of social and cultural conditions that influence the life of an individual or community' or 'the circumstances, objects, or conditions by which one is surrounded'.[2] Within the context of this broader definition of the environment – the one in which individuals and communities both experience and respond to change – climate change is one threat among many. It is a threat that interacts with other processes to create winners and losers (Leichenko and O'Brien, 2008). It is a process that can create shocks or gradual transformations that interact with other processes or events, resulting in non-linear outcomes. A small degree of climate change can push the most vulnerable over the edge into situations of insecurity. Others may be able to cope with or adapt to significant changes in climate, and some may even benefit.

There are many examples of ongoing climate-related catastrophes that have led to beneficial outcomes for those with power and economic resources. For example, the Asian tsunami in 2004 has been described as a 'shock land reform process' where 'nature' took livelihoods away from the very poor while creating space for further recreational businesses in prime coastal areas (Wong, 2009). The impacts of disasters related to climate variability mesh with the specific political and socioeconomic context where it occurs, often deepening inequalities and expanding social exclusions. When tropical cyclone Nargis hit Myanmar (Burma) in May 2008, it left behind 100 000 dead people. The coasts of Burma are highly vulnerable to flooding and storms surges, yet vulnerability and insecurity of Myanmar's inhabitants in relation to climate variability cannot be separated from the oppressive political system that rules the lives of the Burmese. The mixture of political violence and increased climate variability makes the experiences and possibilities for mitigation and adaptation of people in that region particularly difficult (Seekins, 2009). An explicitly normative and broader interpretation of climate change as an issue of human security may be more effective in addressing the complex interactions between a dictatorial political system, ongoing situations of chronic poverty and exclusion, lack of voice and climate variability.

A focus on climate change, ethics and human security creates an opening for new visions and alternatives based on explicitly normative framings of climate change. One conception of human security that embraces an explicitly normative perspective has been proposed by the UN Commission on Human Security (2003). Building on Amartya Sen's critique of welfare economics and associated methods, such as cost–benefit analysis, the Commission considers human security as the protection of the vital core of all human lives to enhance human fulfilment:

---

[2] Merriam-Webster Dictionary: http://www.merriam-webster.com/dictionary/environment

Human security means protecting fundamental freedoms – freedoms that are the essence of life. It means protecting people from critical (severe) and pervasive (widespread) threats and situations. It means using processes that build on people's strengths and aspirations. It means creating political, social, environmental, economic, military and cultural systems that together give people the building blocks of survival, livelihood and dignity.

(Commission on Human Security, 2003: 4)

Although this definition does not specifically address environmental matters, it can potentially reframe the issue of climate change in a way that is more attuned to the goals of global justice. This opens a space for discussion of the ways in which addressing climate change goes hand in hand with an open and reflexive critique of ideas about modernisation, development and quality of life that have led to the climate crisis in the first place, and to discussions of the role of alternative visions about the meaning of development (Jackson, 2009). Moreover, the linking of ethical and justice arguments about climate change with the concept of human security also opens a space for a revised and perhaps much more policy- and action-oriented version of the ethical arguments that have appeared in the literature thus far. In addition to a clear normative component, one advantage of using human security to frame environmental issues is that it links together different issues and allows one to look at power, politics and the contextual factors that create insecurities. For example, dominant conceptualisations of development may be part of the root causes of poverty and destitution, particularly in relation to race, class and gender (Lawson and St.Clair, 2009).

The Human Security Network, a coalition of nations established in 1999 to bring international attention to new and emerging issues of human security, has identified a number of areas for collective action (Human Security Network, 2009). Some of the topics, themes or key threats that have been addressed by the Human Security Network include land mines, HIV/AIDS, globalisation, trafficking in humans and, more recently, climate change. Climate change is a latecomer to the human security agenda, largely because it has not been widely framed as issues of human security in research and policy debates. The focus has been on how humans affect the environment, more than on what environmental change means for individuals and communities who are faced with the interacting consequences of multiple global change processes. Ironically, the recent discourse on security and climate change emerging amid concerns about conflicts and migration threatens to supplant discussions about 'human' security (see Dalby, 2008).

As with any concept, human security has its strengths and weaknesses, and proponents and detractors. Human security has been criticised as being vague and unfocused, and for saying nothing different from human development and other such concepts. It has also been accused of militarising human development. These criticisms, which have been addressed elsewhere (see Gasper, 2005; Matthews *et al.*,

2010), underestimate the importance of human security as an emerging discourse that places individuals and communities at the centre of the analysis. It draws attention to increasing inequities in the world and the processes that create them. It raises ethical questions regarding rights and responsibilities. Finally, it forces a rethinking of current political systems, as well as a questioning of the relevance of existing social contracts (see Hayward and O'Brien, Chapter 11).

Framed as an issue of human security, climate change moves out of the 'environment' box to occupy a much bigger arena. Strategies for addressing climate change consequently must extend beyond government departments and ministries – especially beyond ministries of environment – and become part of more complex political, social and public discussions about what type of a future we want and how it can be achieved (see Bizikova *et al.*, Chapter 9). Such discussions are not unproblematic, for they raise the question of who the 'we' is that decides. There is certainly a need to move beyond individualistic perspectives of human security and beyond western ethical thinking. There is also a need to interrogate the foundations of the dichotomies that permeate many world views, leading to 'us versus them' and 'me versus others' perspectives. Tensions among perceptions of individual and collective human security are reflected in the decades-long discourse on economic development, growth, markets, industrialisation and identification of economic progress and the good life, and the veritable fact that most people in the world, including many in advanced economies, live amid insecurity. Such perspectives become increasingly untenable as climate change affects both individual and collective wellbeing. As Kirby (2006: 151) argues, 'The disjuncture between individual and collective destinies is one of the greatest dramas of our time.'

Contributions to this book

The chapters in this book come from an interdisciplinary group of scholars, including social scientists and philosophers, who seek to draw connections between issues of human security, equity, justice, fairness, ethics and responsibility in relation to climate change. The chapters are structured according to the main themes discussed in this introduction: framings, equity, ethics and reflexivity. All chapters share a holistic perspective on the questions posed by climate change, including its relation to other threats to human security, and to poverty and development.

The first theme of the book emphasises why framings matter from the perspective of human security. The two chapters by Des Gasper and Jon Barnett consider how human security framings can (re)direct the attention of those engaged in climate change debates towards broader, more holistic and integrated understandings of the notion of security. In Chapter 2 on 'The idea of human security', Gasper discusses human security as an emerging concept. He argues that the term 'human security' should be placed at the centre of policy and intellectual debates about global challenges. As an intellectual framework, it can provide a shared language for a global normative commitment. Gasper argues that both actors and functions of the discourse should be broadened to include more than states and multilateral agencies. The human security discourse, he suggests, can serve as an idiom that plays important roles in motivating and directing attention, and in problem recognition, diagnosis, evaluation and response. It motivates action in certain directions through the types of values that it highlights, emphasising, for example, system redesign rather than simply palliative measures to reduce crises.

Jon Barnett's chapter on 'Climate change science and policy, as if people mattered' explains why climate change research must include a deeper understanding of the human context, including cultural perspectives. Building on his research on climate change impacts, vulnerability and adaptation in the South Pacific, Barnett argues that the ideas and methods of dominant scientific models present impediments and limitations for addressing climate change. He discusses why a human security framing is not only a necessity, but a moral imperative for understanding what climate change means to Pacific Islanders. The current predominance of Integrated Assessment Models (IAMs) over bottom-up approaches to vulnerability and adaptation assessments has profound implications for policy and local engagement. A human security framing would contribute to more powerful arguments about potential social losses, and it can humanise risks for decision makers, which may lead to political action on both mitigation and adaptation.

The second theme discussed in the book is equity. In Chapter 4 on 'A "shared vision"? Why inequality should worry us', Timmons Roberts and Brad Parks discuss the many dimensions of inequality related to climate change, and argue that addressing these is a prerequisite for successful international climate negotiations. Shifting the focus from inequities related to greenhouse gas emissions towards inequities in the global distribution of wealth and power is necessary, as inequality dampens cooperative efforts and polarises policy preferences. In short, they suggest that breaking the current North–South stalemate on global climate policy will require unconventional policy interventions, including hybrid approaches that go well beyond issues of financing. Without first building a climate of trust based on 'negotiated justice', Roberts and Parks argue that there is unlikely to be a shared vision for long-term cooperative action.

In Chapter 5 on 'Fair decision making in a new climate of risk', Neil Adger and Don Nelson examine climate change adaptation as a central part of the changing landscape of human security, arguing that it involves justice dilemmas that are closely linked to issues of sustainability and equitable development. Access to decision making and participation in planning for the future under new climate risks is considered a key to enhancing human security. This ability to enhance one's own future, both individually and collectively, raises issues of process and procedural fairness. These themes are explored through case studies in Brazil and Tobago, which show that pathways towards adaptation require radical changes to the status quo, including a diversification of legitimate knowledge and interests in processes of adaptation. Adaptation, they argue, is about creating the conditions for response throughout society as a whole, which requires recognising and addressing power imbalances rather than reinforcing existing inequities and the perpetuation of narrow interests.

Ethics is the third theme contributions to this volume present. The three chapters by Desmond McNeill, Simon Caney and Stephen Gardiner discuss different understandings and arguments about the relations between ethical thinking and climate change. All, however, share a dissatisfaction with the market-driven and economistic conceptions of value and ethics. In Chapter 6 on 'Ethics, politics, economics and the global environment' McNeill argues that climate change is at the same time a political, an economic and a moral problem. He recognises that market-based economic instruments, including taxes and subsidies, are indeed very powerful; and, in the right hands, they have the potential to contribute to the massive behavioural changes that are required to meet the challenges of climate change. However, to allow the market alone to determine how resources are to be allocated is not simply to risk inequitable and inefficient outcomes; it is to abrogate moral responsibility. Standard economic analysis frames issues such as climate change in a way that excludes ethical and political dimensions, and tends rather to conceal the social inequity and environmental costs that climate change brings. Recognising that the unprecedented challenge of climate change is not simply a technical issue, but a moral issue, McNeill questions the dominant position that economics enjoys in relation to other social sciences, particularly when it comes to advising policy makers. He warns that although economics is part of the solution, it is also part of the problem.

In Chapter 7, Simon Caney brings us to a thorough analysis of the relations between human rights and climate change. In 'Human rights, climate change and discounting' argues that climate change jeopardises human rights, and offers solid arguments as to why this jeopardy is sufficient to impose obligations on others. He establishes that people have the right not to be exposed to dangerous climate change. After having established this right, Caney examines the way in which dominant literature on the economics of climate change discounts fundamental rights. The prevailing intellectual framework employed to analyse climate change by policy elites is cost–benefit analysis, but Caney shows that there is more to political morality than this. The kinds of considerations that we normally invoke to defend human rights, Caney maintains, entail that persons have a human right not to suffer from the ill effects of global climate change. Climate change undermines persons' human rights to a decent standard of health, to economic necessities and to subsistence. Caney clearly states that the rights of a person in the twenty-third century have the same moral standing as the rights of a person in the twenty-first century. Even if this chapter does not present a fully elaborated theory of justice, it serves to establish the case for a human rights-based approach to climate change.

In Chapter 8 on 'Climate change as a global test for contemporary political institutions and theories', Stephen Gardiner addresses a different yet equally fundamental aspect of ethics, namely our capacity to interrogate the failures of philosophical systems used by policy makers. To answer the question of why political philosophy should be concerned about global environmental change, in general, and climate change, in particular, Gardiner proposes a minimal global test for social and political institutions and theories. He argues that conventional versions of both are failing to deal with climate change, or what he calls 'the perfect moral storm'. He then explains why existing global systems are poorly placed to handle such scenarios and draws attention to humanity's initial and unsuccessful response to the climate crisis to confirm his conjecture. He identifies a number of weaknesses in moral and political theory, and illustrates his concerns by focusing on one particular kind of moral and political theory, utilitarianism. In particular, he addresses the most influential version of utilitarianism with respect to climate change, namely the use of cost–benefit analysis within a conventional economic framework. He points to the inadequacies of these and other theories, and proposes shifting the focus away from the prevailing economic paradigm for evaluating institutions. Instead of worrying about how to maximise or optimise overall benefits, understood in market terms, the core concern might be with securing central goods, such as human rights, basic needs and capabilities. Passing the global test, he concludes, requires, among other things, a shift from complacency.

The last set of chapters in the volume address the theme of reflexivity, which includes the capacity to respond to long-term threats to human security. Chapter 9 on 'Linking sustainable development with climate change adaptation and mitigation', by Livia Bizikova, Sarah Burch, Stewart Cohen and John Robinson, explores the capacity to respond to climate change and create a sustainable future. Moving away from traditional integrated assessments, plagued by the lack of thorough integration of social and institutional domains (a necessary condition for facilitating decision making under conditions of uncertainty), the chapter discusses a participatory integrated assessment (PIA) framework. The PIA, the authors maintain, can be used as a platform for organising research, providing an ongoing learning opportunity for both researchers and practitioner/stakeholder partners. Within the PIA, scenario and backcasting tools can be used in conjunction with other case-specific methods (e.g. from forestry, water management and urban planning), as well as dialogue support methods, such as visualisation and decision-support models. These methods, which explicitly consider tradeoffs and synergies between adaptation and mitigation, can be considered key tools for identifying sustainable futures. A 'learning by planning' stakeholder-driven approach creates linkages between information produced by scientists and information used in decision making. It links climate change and sustainable development, generating integrated scenarios that respond to people's actual needs and visions for the future.

In Chapter 10 on 'Global poverty and climate change: towards the responsibility to protect', Asunción Lera St.Clair critiques the lack of reflexivity that dominates emerging discourses on climate change and poverty reduction. She argues that the most efficient way to promote sustainable responses to climate change is the immediate eradication of severe poverty, combined with the development of solid welfare systems for social protection, and a minimisation of inequalities. This clearly involves radical changes in the theory and practice of poverty reduction, including a shift away from the emerging focus on market-based solutions to climate change. It also involves greater attention to the concept of 'responsibility', including the responsibility to protect vulnerable populations. Allocating responsibilities for climate change starts with democratic deliberation, with open-ended, constant and ongoing public and political debate. St.Clair argues that these are justified not only because of the intrinsic value and dignity of all human beings, but because they also have a fundamental instrumental value – namely building solidarity and global social cohesion.

In Chapter 11 on 'Social contracts in a changing climate: security of what and for whom?', Bronwyn Hayward and Karen O'Brien critically reflect on calls for new social or environmental contracts in response to climate change. Social contracts play an important role in defining the reciprocal rights and responsibilities of the state and citizens, as well as of citizens to each other. However, social contract theory, and the market liberal values embedded in it, has become problematic over time. Social contract theory has often served as a legitimating tool for power relationships that have perpetuated human injustice, unsustainable resource extraction and colonialisation. The problems with contract thinking are exacerbated when new political solutions are subject to ineffective public scrutiny or debate. Hayward and O'Brien argue that it is unclear what forms of security the new climate contracts will provide, and for whom. New climate treaties, 'green new deals' and carbon contracts may appear to be elegant political solutions but, in reality, they may also serve to obscure the complex underlying processes that have contributed to climate change, displacing or exacerbating environmental injustice over time and space. Drawing inspiration from an alternative vision of a social contract in the writings of Rousseau and the critical thinking of contemporary theorists, the authors propose a meshed solution of local compacts or political agreements developed within a framework of interlinked local, regional and international review and robust public debate.

In the concluding chapter, we take up some of the key challenges in operationalising and 'securing' human security in the near future. In particular, we discuss how human security perspectives can contribute to a shift in the discourse and to the framing of a new science on climate change. We review critiques of the institutionalised discourse on climate change and then discuss what a 'new science' on climate change might look like. We emphasise that it should be able to integrate insights from the social sciences, humanities and other fields with emerging findings on how human activities influence the Earth System, and it should place issues of justice, ethics, responsibility and human security at the forefront of policy debates. We present examples of the questions that this new science might raise and discuss some key themes related to the social context, the institutional context and the human context. Finally, we describe how the concept of human security might be translated into a global perspective. We recognise that much work remains to be done to integrate the beliefs, values and aspirations of diverse individuals and communities into a coherent vision for the future, particularly if this vision is to include global-scale responses that prioritise and create human security. However, the key argument is that the knowledge that is needed to inform both discussions and actions on climate change needs to be broadened. Many of the questions raised by climate change have no clear scientific answers, and normative and ethical considerations need to play a far more visible role in climate change debates. Framing climate change differently, as we argue in this book, can provide new ways of both seeing and resolving what is undoubtedly one of the biggest and most complex challenges to human security.

References

Adger, W. N., Paavola, J., Huq, S. and Mace, M. J., eds. 2006. *Fairness in Adaptation to Climate Change*. Cambridge, MA: MIT Press.

Barnett, J. 2001a. Adapting to climate change in Pacific Island countries: the problem of uncertainty. *World Development*, **29**(6), 977–93.

Barnett, J. 2001b. *The Meaning of Environmental Security: Ecological Politics and Policy in the New Security Era*. London: Zed Books.

Beck, U., Giddens, A. and Lash, S. 1994. *Reflexive Modernization: Politics, Tradition and Aesthetics in the Modern Social Order*. Stanford, CA: Stanford University Press.

Berkes, F. 2008. *Sacred Ecology*. London: Routledge.

Cambridge Programme for Sustainability Leadership 2009. *Video: One Degree War Plan*. Cape Town: CPSL. Available online: http://www.cpsl.co.za/2009/05/video-one-degree-war-plan/.

Castree, N. 2005. *Nature*. London: Routledge.

Commission on Human Security 2003. *Human Security Now*. New York: Commission on Human Security. Available online: http://www.humansecurity-chs.org/finalreport/English/FinalReport.pdf.

Dalby, S. 2002. *Environmental Security*. Minneapolis, MN: University of Minnesota Press.

Dalby, S. 2008. Geographies of environmental security. In G. Youngs and E. Kofman, eds., *Globalization: Theory and Practices*, 3rd edn. London and New York: Continuum, pp. 29–39.

Dalby, S. 2009. *Security and Environmental Change*. Cambridge, UK: Polity Press.

Devall, B. and Sessions, G. 1985. *Deep Ecology: Living as if Nature Mattered*. Layton, UT: Gibbs Smith.

Esbjörn-Hargens, S. and Zimmerman, M. E. 2009. *Integral Ecology: Uniting Multiple Perspectives on the Natural World*. Boston, MA: Integral Books.

Folke, C. 2006. Resilience: the emergence of a perspective for social–ecological systems analyses. *Global Environmental Change*, **16**, 253–67.

Forsyth, T. 2003. *Critical Political Ecology: The Politics of Environmental Science*. London: Routledge.

Freire, P. 1970. *Pedagogy of the Oppressed*. Translated by Myra Bergman Ramos. New York, NY: Continuum.

Gasper, D. 2005. Securing humanity: situating human security as concept and discourse. *Journal of Human Development*, **7**(2), 221–45.

GECHS 1999. *Science Plan: Global Environmental Change and Human Security*. Bonn: International Human Dimensions Programme (IHDP).

Global Humanitarian Forum 2009. *The Anatomy of a Silent Crisis*. Geneva: Global Humanitarian Forum.

Gore, A. 2006. *An Inconvenient Truth: The Planetary Emergency of Global Warming and What We Can Do About It*. Emmaus, PA: Rodale.

Gunderson, L. and Holling, C. S., eds. 2002. *Panarchy: Understanding Transformations in Human and Natural Systems*. Washington, DC: Island Press.

Harding, S. 2006. *Animate Earth: Science, Intuition, and Gaia*. White River Junction, VT: Chelsea Green Publishing Company.

Hulme, M. 2009. *Why We Disagree About Climate Change: Understanding Controversy, Inaction and Opportunity*. Cambridge, UK: Cambridge University Press.

Human Security Network 2009. The Human Security Network. http://www. humansecuritynetwork.org.

IPCC 2007. *Climate Change 2007: The Physical Science Basis*. Contribution of Working Group I to the Fourth Assessment Report of the Intergovernmental Panel on Climate Change (IPCC). Cambridge, UK: Cambridge University Press.

Jackson, T. 2009. *Prosperity Without Growth? The Transition to a Sustainable Economy*. London: UK Sustainability Commission.

Jasanoff, S. and Martello, M. L., eds. 2003. *Earthly Politics: Local and Global in Environmental Governance*. Cambridge, MA: MIT Press.

Jasanoff, S. and Wynne, B. 1998. Science and decision making. In S. Rayner and E. Malone, eds., *Human Choice and Climate Change*. Columbus, OH: Battelle Press, pp. 1–87.

Kegan, R. 1994. *In Over Our Heads: The Mental Demands of Modern Life*. Cambridge, MA: Harvard University Press.

Kirby, P. 2006. *Vulnerability and Violence: The Impact of Globalization*. London: Pluto Press.

Klein, R. J. T., Eriksen, S. E. H., Næss, L. O. *et al.* 2007. Portfolio screening to support the mainstreaming of adaptation to climate change into development assistance. *Climatic Change*, **84**(1), 23–44.

Lawson, V. and St.Clair, A. L. 2009. The relevance of critical global poverty studies in the re-framing of environmental change as an issue of human security. *IHDP Update*, 2009/2, 25–9.

Leichenko, R. M. and O'Brien, K. L. 2008. *Environmental Change and Globalization: Double Exposures*. New York, NY: Oxford University Press.

Lovelock, J. 2009. *The Vanishing Face of Gaia*. United Kingdom: Penguin.

Maniates, M. 2002. Individualization: plant a tree, buy a bike, save the world? In T. Princen et al., eds., *Confronting Consumption*. Cambridge, MA: MIT Press, pp. 43–66.

Martello, M. L. and Jasanoff, S. 2004. Introduction: globalization and environmental governance. In S. Jasanoff, and M. L. Martello, eds., *Earthly Politics: Local and Global in Environmental Governance*. Cambridge, MA: MIT Press, pp. 1–30.

Matthews, R. A., Barnett, J., McDonald, B. and O'Brien, K. L., eds. 2010. *Global Environmental Change and Human Security*. Cambridge, MA: MIT Press.

McNeill, D. and St.Clair, A. L. 2009. *Global Poverty, Ethics and Human Rights: The Role of Multilateral Organisations*. London: Routledge.

Millennium Ecosystem Assessment 2005. *Ecosystems and Human Well-being: Synthesis*. Washington, DC: Island Press.

Miller, C. and Edwards, P. 2001. *Changing the Atmosphere: Expert Knowledge and Environmental Governance*. Cambridge, MA: MIT Press.

Müller, B. 2002. *Equity in Climate Change: The Great Divide*. Oxford, UK: Oxford Institute for Energy Studies, with support of Shell Foundation.

Naess, A. and Rothenberg, D. 1993. *Ecology, Community and Lifestyle: Outline of an Ecosophy*. Cambridge, UK: Cambridge University Press.

O'Brien, K. and Leichenko, R. 2003. Winners and losers in the context of global change. *Annals of the Association of American Geographers*, **93**(1), 99–113.

O'Brien, K. and Leichenko, R. 2006. Climate change, equity and human security. *Die Erde*, **137**(3), 223–40.

O'Brien, K., Quinlan, T. and Ziervogel, G. 2009. Vulnerability interventions in the context of multiple stressors: lessons from the Southern Africa Vulnerability Initiative (SAVI). *Environmental Science and Policy*, **12**(1), 23–32.

Parry, M., Cancioni, O., Palutikof, J., van der Linden, P. and Hanson, C., eds. 2007. *Climate Change 2007: Impacts, Vulnerability, and Adaptation*, Report of Working Group II to the Fourth Assessment Report of the Intergovernmental Panel on Climate Change (IPCC). Cambridge, UK: Cambridge University Press.

Parry, M., Lowe, J. and Hanson, C. 2009. Overshoot, adapt and recover. *Nature*, **458** (30),1102–103.

Pogge, T. 2004. The first United Nations Millennium Development Goal: a cause for celebration? *Journal of Human Development*, **5**(3), 377–97.

Roberts, T. J. and Parks, B. C. 2006. *A Climate of Injustice: Global Inequality, North–South Politics, and Climate Policy*. Cambridge, MA: MIT Press.

Sen, A. 1999. *Development as Freedom*. New York, NY: Anchor Books.

Seekins, D. M. 2009. State, society and natural disaster: Cyclone Nargis in Myanmar (Burma), *Asian Journal of Social Science*, **37**(5), 717–37.

St.Clair, A. L. 2006a. The World Bank as a transnational expertised institution. *Journal of Global Governance*, **12**(1), 77–95.

St.Clair, A. L. 2006b. Global poverty: the co-production of knowledge and politics. *Journal of Global Social Policy*, **6**(1), 57–77.

St.Clair, A. L. 2006c. Global poverty: development ethics meets global justice. *Globalizations*, **3**(2), 139–58.

Steffen, W. A., Sanderson, P. D., Tyson, J. *et al.* 2004. *Global Change and the Earth System: A Planet Under Pressure*. Berlin: Springer.

Stern, N. 2007. *Stern Review on the Economics of Climate Change*. Cambridge, UK: Cambridge University Press.

United Nations 2005. *The Inequality Predicament: Report on the World Social Situation 2005*. New York: United Nations Department of Economic and Social Affairs.

United Nations Development Programme (UNDP) 1994. *Human Development Report 1994*. New York, NY: Oxford University Press.

United Nations Development Programme (UNDP) 2007/2008. *Fighting Climate Change: Human Solidarity in a Divided World*. New York: United Nations Development Programme.

Wong, P. P. 2009. Rethinking post-tsunami integrated coastal management for Asia-Pacific. *Ocean & Coastal Management*, **52**(7), 405–10.

Young, O. R., King, L. A. and Schroeder, H., eds. 2008. *Institutions and Environmental Change*. Cambridge, MA: MIT Press.

2

The idea of human security

DES GASPER

Prelude: The surprising spread of 'human security' discourse

The language of 'human security' that became prominent in the 1990s has encountered criticism from many sides. Nonetheless, over the past twenty years it has continued to gain momentum. One encounters it frequently now, not only in debates about physical security, but in discussions of environment, migration, socioeconomic rights, culture, gender and more. Werthes and Debiel (2006: 8) propose that "human security provides a powerful 'political leitmotif' for particular states and multilateral actors by fulfilling selected functions in the process of agenda-setting, decision-making and implementation." I suggest that in order to understand human security discourse and its spread, this specification of actors and functions should be broadened. The relevant actors include more than states and multilateral agencies, and what was originally primarily a language in United Nations circles is now far more encompassing. Like the sister idea of human rights, human security is becoming an idiom that plays important roles in motivating and directing attention, and in problem recognition, diagnosis, evaluation and response.

The concept of 'security' in a human context

The concept of human security redirects attention in discussions of security, beyond the nation-state level, beyond physical violence as the only relevant threat/vector, and beyond physical harm as the only relevant damage. Scores of specific proposed definitions exist.[1] In an earlier study (Gasper, 2005), I organised a range of definitions in an analytical table, which Table 2.1 now extends (the entries in italics indicate diverse possible definitions of human security).

[1] See e.g. http://www.gdrc.org/sustdev/husec/Definitions.pdf or the report of the Commission on Human Security (CHS, 2003).

Climate Change, Ethics and Human Security, eds. Karen O'Brien, Asunción Lera St.Clair and Berit Kristoffersen. Published by Cambridge University Press. © Cambridge University Press 2010.

Table 2.1 *Alternative definitions of human security (HS)*

	Valued Capabilities Expansion (e.g. UNDP 1990)	Human Development in terms of UNDP's longer list of goods (e.g. UNDP 1996)	Basic Needs Only (in terms of types and level)	Life preservation (against structural, not only physical violence)	Personal Physical Security Only (& civil rights)
Attention to *Level* of some Valued Variables (snapshot or trend)	Sen's Capability Approach in minimal form (Sen, 1993)	Human Development Reports' focus (includes physical security)		Picciotto et al. (2006, 2007)	Canadian and Norwegian government definition of HS
HS in terms of *Stability*			'Downturn with stability' (of basic needs fulfilment for everyone)[a]		
HS in terms of *both* Level and Stability		Haq's (1999) maximal definition of HS; Govt. of Japan definition (1999)	Alkire's and Ogata-Sen's definition (CHs, 2003)		

[a] 'Downturn with stability' is a phrase used by Sen (e.g. in Commission for Human Security, 2003), to refer to still maintaining stability of basic needs fulfilment for everyone.

'Human security' is discussed at different scales and with reference to threats of varying scope. Moving through from broader to narrower definitions, it can be treated as the security of the human species, or as the security of human individuals. Within the latter, it may focus on severe, priority threats to individuals, as judged for example by mortality impacts or by the degree of felt disquiet. The severe priority threats are sometimes limited to concern only 'freedom from want' and 'freedom from fear,' or even only the latter. More narrowly still, some authors wish to consider only threats to individuals brought about through violence. Finally, the narrowest conception of human security refers only to threats to physical survival brought about through organised intentional violence (MacFarlane and Khong, 2006: 245–7).

Before further considering the alternative formulations of human security, we should reflect on the concept of security and the significance of the term 'human.' We should be aware of the underlying arguments for claiming and justifying priority, in relation to the choices of what is to be considered as within the concept of human security.

Objective/subjective. The 'security' concept began as a subjective concept, from classical Rome, suggested Wolfers (1962). A subjective security concept must cover the range of whatever concerns are felt as threats (Hough, 2005). So too must an objective security concept, insofar as feelings typically correspond to real possibilities, though they are often based on misunderstandings about probabilities. Objective security can still be distinguished from subjectively felt security, given the poor correlation of their magnitudes which represents one of the core paradoxes of security. For example, Latvia's Human Development Report on human security noted that the Latvian language employs distinct terms for the two concepts. The reported priority threats felt subjectively by Latvians are easily understood but not necessarily predictable: inability to pay for major medical care and old age; and fear of physical abuse at home and of abuse by officials, such as the police (UNDP, 2003).

Means/ends. Two further categories are important. One concerns the means that are intended to achieve safety or the feeling of safety. The experience of not feeling safe from the state bodies that are supposed to promote security and felt safety – a second classic paradox of security – led women in Hamber *et al.*'s (2006) studies to make statements like: "For me the word security in Arabic is not to be afraid. First, not to be afraid to be hungry, to move, to think, and to be misjudged"; "[Security is] . . . not being afraid, and that can be of physical violence but also feeling you have the right to do the things you want to do and say"; and even to a positing of 'security' as a man's word and 'safety' as a woman's word. The Bangladesh Human Development Report on human security found similarly that poor people felt less secure thanks to the police (UNDP, 2002).

The other necessary additional category concerns *being able* to be safe. The Global Environmental Change and Human Security (GECHS) project considers human security as the capacity of individuals and communities to respond to threats to social, human and environmental rights (GECHS, 1999). This formulation leaves people with the responsibility to use that capacity, gives recognition to communities, and gives them space to prioritise threats.

Claiming priority. Security claims are claims of existential threat (Buzan et al., 1998) meant to justify priority responses, including overriding of other claims or rights. Attempts to limit such prioritisation to one type of threat (such as threats of physical damage from violence) and/or one type of referent or target (such as the state) are artificial. The root and usages of the term 'security' do not validate such a restriction. According to Rothschild (1995), for centuries the term applied only to individuals. More recently, the UNDP's 1993 and 1994 Human Development Reports led by Mahbub ul Haq reestablished a broad meaning for human security, in terms of a range of types of threat. This definition followed from suggestions by, for example, Juan Somavia and others in the South American Peace Commission in the 1980s, Lincoln Chen and Ken Booth at the start of the 1990s; and Johan Galtung, Kenneth Boulding and others in peace research a generation earlier (Bilgin, 2003). Some formulations now go so far as to discuss human security in terms of all threats to internationally ratified human rights, though this can weaken the prioritising thrust and has to be balanced by the next idea, that of basic thresholds.

Justifying priority: normative thresholds and understanding the human referent for security. One must not merely claim priority, but also have a plausible basis for it. Some of the debates on human security consider at length the concept of 'security,' yet pay little attention to the content of 'human', as if that has no relevance to interpreting the scope of 'human security'. To mention the individual as one referent for the concept of security is not enough. Attention is required to the nature of the referent.

Beyond the mere fact that humans are embodied persons, being human has various specific requirements. These include partly socially-specific *normative thresholds*, across a range of needs, i.e., minimum levels required for normative acceptability. 'Human security' issues in the area of health, for example, do not include all health issues, only those up to a minimum normatively set threshold (even though that is to some degree historically and often societally specific – see, for example, Owen, 2005; Gasper, 2005). Lack of the threshold distinction leads to an argument for excluding whole issue areas, such as health, from the remit of 'security,' mistakenly believing that this is necessary in order to allow meaningful prioritisation (see MacFarlane and Khong, 2006).

Justifying priority: interconnection, nexuses and tipping points. A typical aspect of justifying priority is to identify a major causal connection from fulfilment or non-fulfilment of the highlighted factor, to a qualitatively different set of other things that have clear normative importance. This is the notion of a *nexus*, a major connection, at least in some situations, between different 'spheres' – for example between environment and peace or war – and thus from one thing to many others. The discourse of insecurity often proposes a particular type of connection: a *causal threshold, flashpoint or tipping point*, or a stress level beyond which dramatic escalation of negative effects occurs, bringing even collapse. For example, beyond certain levels and combinations of stress factors, drastically increased damage occurs to human health, including life expectancy; some combinations bring premature death. Violent death scenarios, let alone violent deaths intentionally promoted by others, are only one type of premature death scenario. Suicides by heavily-indebted farmers have become frequent in parts of India, for example. Arguably, whole societies too can go over a stress tipping point.

Structural limits are central to human security analysis. Beyond the limits, things snap. The 'weak sustainability' hope in environmental economics is inapplicable outside certain bounds; less environmental capital cannot always be substituted for by having more of another type: human, social, or human-built physical capital. Destabilisation of the Earth's regenerative and climate cycles cannot be compensated for by more of other capital types.

To review, 'security issues' concern risks of being or falling below minimum normative thresholds. Security means 'holding on' or 'holding firm', to core values. Especially serious are cases with significant possibilities of collapse; yet while a famine where a social system has collapsed is a prime example of lack of human security, chronic capacity-sapping malnutrition is an example too. Normative thresholds and causal thresholds can be connected; for when a normative threshold is breached a person may erupt, against others or herself, or collapse.

Justifying priority: issues of responsibility and intentionality. Should we consider all matters that involve threats to basic values as human security issues, or only those which are intentionally caused and which are not the victim's own responsibility (thus excluding, for example, smoking-related disease)? Matters which are victims' own responsibility are in fact already excluded by a focus on *capability to be safe*. MacFarlane and Khong's (2006) definition – threats to our physical survival caused by intentional organized violence – goes further and excludes unintentional damage. Their definition is still a human security conception, since it concerns threats to individuals, but is nonetheless very narrow. It excludes climate change from our purview, not only because the threats are not (all) related to physical violence, but because there is no conscious perpetrator of harm. We return to their choice later, and suggest that it mistakes short-term policy convenience for analytical power and long-term relevance.

Security as a visceral concept. Security is not just a prioritising, claiming concept. The way that humans have evolved, the way our consciousnesses are structured, means that some events and things disturb us or destabilise us. Combined with 'human,' 'security' conveys a visceral, lived feel, connecting to people's fears and feelings or to an observer's fears and feelings about others' lives. 'Human security' thus evokes a sense of real lives and persons. Like 'rights,' it touches something deep in our awareness. Part of this may reflect a human priority to avoid losses more than to make gains. Losses can refer to the loss of meaning and identity, and not merely to the loss of things.

Human security as an integrative concept. 'Human security' captures what some other concepts cover, and goes further. Like basic needs analysis it gives substance to the language of 'development,' a language to talk about significant change that does not yet tell us anything about the contents of that significance. It then adds to what basic needs analysis conveys, by for example its stronger link to feelings (Gasper, 2005). It helps to give a sense of direction and priority too within the language of rights, which is about the form of a priority claim but not necessarily about its content or rationale, and which otherwise can bring an absolutization of the convenience and property of the powerful (Gasper, 2007).

The human security concept thus concerns an assurance for individuals (and societies, and the species) of normatively basic threshold levels in priority areas. It connects a series of ideas: objective and subjectively felt security; normative priorities for what it is to be human, including a sense of meaning and identification; causal nexuses, tipping points, and awareness of possibilities of collapse. We thus see that there is a discourse of 'human security' that goes beyond just a single concept. Indeed, if we highlight different choices of inclusions and emphases, it is possible to distinguish a family of related but varied discourses.

Components of the 'human security' discourse(s)

In an earlier paper I examined 'human security', in particular the UNDP human security approach, as a discourse that employs the concept and label, but includes more (Gasper, 2005). Elements of this UN discourse were specified as follows. The first four elements are shared with UNDP's sister discourse of human development:-

- A heightened normative focus on individuals' lives.
- More specifically, a focus on reasoned freedoms, the ability of persons and groups of persons to achieve outcomes that they have reason to value.
- 'Joined-up thinking' (Gasper and Truong, 2005) that looks at the interconnections between conventionally separated spheres (notably, different polities; and within polity-economy-society-ecosystems), and not least at the nexus between freedoms from want and indignity and freedom from fear. Correspondingly it tries to build policy coherence across conventionally separated spheres.
- A global span normatively as well as for explanatory purposes; covering all persons, worldwide, as in human rights discourse.

Human security discourse adds at least three elements, which contribute to a stronger motivational basis than in the original UN human development approach.

These elements help to mobilise attention and concern and to sustain a global normative commitment or 'joined-up feeling.' They include the following:

- A focus on basic needs.
- More specifically, an insistence on basic rights for all. This strengthens the focus on individuals, compared to the human needs and human development traditions.
- A concern for the stability as well as the average levels of important freedoms.

This notion of 'human security' is a complex package, perhaps too much so for MacFarlane and Khong (2006), the international relations specialists who were commissioned to discuss the notion for the UN Intellectual History Project. They miss the basic needs point about minimum required levels, which differentiates human security work from the pure human development approach. Likewise, they suggest wrongly that the Commission on Human Security's report (CHS, 2003) was concerned only with assuring stability of fulfilment, not primarily with assuring basic levels.[2]

Let us examine more fully the various elements and how they fit together. The first heading below relates especially to what O'Brien and Leichenko (2007) call the equity dimension in human security thinking. The next two headings relate to what they call the connectivity dimensions.

Humanism: integrating the international 'human' discourses. Human security work synthesises ideas from the preceding 'human discourses' of human development, human needs, and human rights (Gasper, 2007). As the UN Intellectual History Project highlights, human rights language gave an independent value status to prioritised individual freedoms, and a universal scope of consideration. It implied obligations on states to meet these priorities, and implied legitimate recourse by persons without those rights, to hold states accountable (Jolly *et al.*, 2004: 187). To supplement this, 'the human development approach introduces the idea of scarcity of resources, the need to establish priorities, and sequencing of achievement in the promotion of human rights' (Jolly *et al.*, 2004: 177). Human security language combines the human rights insistence on the importance of each individual, with a human development insistence on priority sequencing given the scarcity of resources.

The heightened normative focus on individuals' lives gives human security thinking a radical thrust. Picciotto *et al.* (2006, 2007), for example, adopt life-years rather than the Human Development Index as primary performance measure. We should not trade-off extra years of life for people who live only forty years, against an increase in average per capita income. Instead we should take as a priority human right a

---

[2] '... [we make] an examination of the report of the Commission on Human Security, which made a strong case for viewing human security as the protection of individuals from the vulnerabilities associated with sudden economic downturns' (MacFarlane and Khong, 2006: 16).

lifespan of, say, three score years and ten, the natural span that is relatively easily attainable and only with much greater difficulty extendable. It is the lifespan that has been attained and assured at relatively low per capita income in places like China, Costa Rica, Cuba, Jamaica, Kerala and Sri Lanka.

Humanism II: a holistic perspective at the level of the individual. We find in human security work an anthropological type concern for understanding how individual persons live, that provides microfoundations for explanatory macro theory. People seek security, of various sorts: bodily, material, psychological and existential (including via family, friends, esteem, systems of meanings). All of this is long familiar, but regularly forgotten. One recent locus of such understanding has been the basic needs school in conflict studies from the 1970s on (Burton, 1990). Human security thinking has given it a more capacious home. This holistic perspective at individual-level gives a broader (UNDP) perspective on human security decisive advantages over a narrower (Canadian) perspective, let alone the MacFarlane-Khong variant.

Trans- or supra-disciplinary explanatory synthesis: a (selective) holistic approach at the level of larger systems. At supra-individual levels, human security thinking stresses the interaction of economic, political, social, cultural, epidemiological, military and other systems that have conventionally been treated separately in research and policy. This 'joined-up thinking' is holistic in spirit but not totalising in scope; the particular interconnections to be stressed will be selected according to their importance case-by-case. Several interviewees in the UN Intellectual History Project express this holistic spirit:

"the basic premise of the [UN] charter, that you really can't have peace unless the rights of nations great and small are equally respected. . . . [and] the basic premise of the Declaration of Human Rights, that you can't have peace within a country unless the rights of all, great or small, are equally respected."

(Virendra Dayal, quoted by Weiss et al., 2005: 151)

". . . all the conflicts that [some rich governments] are giving rise to in an interdependent world precisely by ignoring the human rights and the democratic principles that they supposedly espouse."

(Lourdes Arizpe, quoted by Weiss et al., 2005: 415)

Juan Somavia, who ran the 1995 Copenhagen summit on social development that took steps down the broader human security path, noted how 'the constitution of the ILO . . . already in 1919, says that peace is linked to social justice', and quoted Pope Paul VI's declaration in 1969 at an ILO conference that "Development is the new dimension of peace" (both cited by Weiss *et al.*, 2005: 299).

Outweighing such ideas though: "The whole system has pushed, pushed, in educational terms, towards specialization, when the reality of the world has been

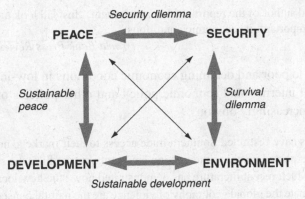

Figure 2.1 The conceptual quartet and six linkages (based on Brauch, 2005)

pushing more and more towards integration" (Somavia, cited on p. 429). Educational narrowing blinds us to interconnection and helps to generate new threats. As Zygmunt Bauman describes, extreme intellectual specialisation – 'close-focusing' of the types done so successfully in science and technology – leads to waves of unforeseen effects when we act on the resulting powerful but narrow knowledge. It has led us into Ulrich Beck's 'Risk Society', where every 'advance' creates new messes and "the line beyond which the risks become totally unmanageable and damages irreparable may be crossed at any moment" (Bauman, 1994: 29).

Figure 2.1 identifies more specifically the interconnections which are meant to justify and be revealed by 'joined-up thinking'. Brauch (2005) presents four trad-itional foci, which imply six types of possible major bilateral interconnection. Though Brauch uses 'security' to mean security against violence (or even only inter-state violence), and his table presents the interconnections in terms of binary relations, each side of each binary relation is linked to all the other foci. The human security research programme posits that in at least some important cases the inter-connections are ramifying and major, and require us to move beyond traditional problem framings.

This holistic spirit has a grand sweep, but are the declared linkages adequately established? The linkage from carbon-based economic growth to global climate change is more than sufficiently demonstrated. Regarding economic performance and conflict, Paul Collier et al.'s 2003 study for the World Bank showed a strong correlation of violent conflict with both poverty and low growth.

By analyzing 52 major civil wars between 1960 and 1999 it found that the common thread was often a poor and declining economy combined with a heavy dependence on exports of natural resources such as diamonds, gold or oil. "Some countries are more prone to civil wars than others but distant history and ethnic tensions are rarely the best explanations,"

Paul Collier, lead author of the report, said in a statement. "Instead look at a nation's recent past and, most important, its economic conditions."

(World Bank Press Review: 15 May 2003)

Next, linkages to poor and declining economic conditions in low-income countries from aspects of international economic policy and other policies of rich countries have become increasingly obvious.

- Rich countries have restricted Southern trade access to their markets, notably in agriculture, and yet expected no consequences: no emigrants, no conflicts, no spillover of stress or suffering. Much recent literature has demonstrated how "the new local wars that have come to dominate the global geography of violence are the natural consequence of formal rules that make the criminal economy of illegal trafficking in drugs, weapons and people far more attractive to poor and marginalised countries than legal economic pursuits" (Picciotto, 2005: 3).
- Rich countries have energetically exported arms and imagined these will not be used. "Most weapon-exporting countries provide export credit guarantees for weapons purchases by developing countries" (Picciotto, 2005: 6).
- Rich countries imposed bone-crunching economic structural adjustment on low income countries and expected no wider consequences. An income shock of −5% raises chances of civil war by 50% (Picciotto *et al.*, 2006). Prior to the 1994 Rwanda genocide, the country faced an income decline of 40% as IMF-imposed adjustment was piled on top of the effects of slump in the world coffee market. The economic impacts of civil wars are themselves so immense (e.g. "In Rwanda, Bosnia and Lebanon GDP fell to 46%, 27% and 24% of the pre-conflict peaks", Picciotto *et al.*, 2006: 6) as to thereby greatly raise the chances of perpetuation of the war.
- As a latest aspect of policy incoherence, international policies on governance have blocked aid to states that are adjudged to not already have good governance, and have thereby undermined international security policy.

Roles

The idea of 'human security' plays various roles: first, it provides a shared language, that highlights and proclaims a new perspective in investigation; second, it guides evaluations, through its emphasis on certain priority performance criteria; third, it guides positive analyses, through its emphases on which outcomes are important to explain and which determinants are legitimate to include; fourth, it similarly focuses attention in policy design, by directing attention to a particular range of outcomes as being important to influence and a particular range of means as being relevant to consider; and fifth, it motivates and inspires action in certain directions, through the types of value which it highlights and the range of types of experience to which it leads us to attend.

In earlier work on human security thinking in or linked to the UN system I have tried to elucidate these roles.[3] The first column of Table 2.2 below summarises the arguments, drawing also on the previous sections of this chapter. Behind the familiar features – a focus on individuals not only on generalized categories such as national income or averages, and a wider scope both of the areas considered under 'security' and of attention to contributory factors – lie the deeper commitments: the motivating concern of 'joined-up feeling', partnered by the holistic vision of wide-ranging attention to human experience and interconnections therein.

A human security research programme in the universities of Marburg and Duisburg in Germany complements this UN-centred research through its investigations of the 'Human Security Network' of Canada, Norway, and ten other countries and of the work of Japan and the European Union.[4] Table 2.2 compares the Marburg–Duisburg work and my picture of components and roles, and gives illustrations and extensions for both specifications. The German work too is organised by a perspective on what are the roles of a human security intellectual framework. It specifies three: (1) explanation and orientation, (2) coordination and action-related decision guidance, and (3) motivation and mobilisation. They correspond to the last three roles I identified. Let us explore some of these roles further.

Unexpected insights and situation-specific understandings. In explanation, the human security approach provides fresh situation-specific understandings and insights, by applying a non-conventional boundary-crossing perspective in ways tailormade to specific cases. Jolly and BasuRay (2007) have reviewed the many national Human Development Reports focused on human security, to test what if anything the perspective adds. The mandate to look broadly at sources of insecurity, but to be selective according to the particular concerns, constellations and connections extant in a particular country, generates unexpected and practical diagnoses and proposals. The analyses are restricted neither by arbitrary *a priori* disciplinary habits in regard to scope, nor by fixed prescriptions or proscriptions from a global centre about what should be included or excluded. Further examples along these lines are found in work that uses a human security approach to consider environmental and climate change, such as by Jon Barnett and Karen O'Brien (in this volume).

Focusing policy design on foundational prevention rather than crisis management. In policy design, a human security perspective emphasises system re-design to reduce chances of crises rather than palliative measures when crises have hit. Lodgaard (2000) argued that:

---

[3] Gasper, 2005, 2007, 2008; Gasper and Truong, 2005.

[4] Coming from a state security/International Relations background, and with a focus on the Human Security Network countries, some of this work underplays the Basic Needs and Human Rights aspects in the UN-Japan line of human security discourse, and mistakenly separates them from physical security – as if physical security is not part of basic needs, and as if one does not fear lack of basic necessities (see Bosold and Werthes, 2005: 86; Werthes and Bosold, 2006: 25)

Table 2.2 *The components and roles of the idea and discourse of human security*

Roles of an idea/discourse	Gasper (2005, 2007, 2008)	Werthes and Bosold (2005, 2006)
1. To provide a shared language	Besides a concept, 'human security' is also: - A discourse, whose elements are asterisked below * *A striking and evocative label*	Within this shared language people can flexibly respond to their own situation and own priorities. Yet it also provides, in overlap areas, a frame 'for concerted policy projects, *par excellence* illustrated in the [Human Security Network]' (Werthes and Bosold, 2006: 23).
2. To guide evaluations of situations and performance	A normative focus on individual lives, viz: * Focus on individuals' reasoned freedoms * *A concern for stability as well as levels*	(a) From state focus to individual focus; this is the first of Werthes and Bosold's four elements of a proposed share core of HS (2006: 25; also Bosold and Werthes, 2005: 99). Policy language provides one instrument for holding its users accountable (Werthes and Bosold, 2006)
2* – Human focus	* JUF: 'Joined-up feeling', for all individuals – this is the spirit of human rights (HRs) discourse Edson: 'human security is about protecting the common good' (2001: 84)	(b) 'People should have the opportunity to live decently and without threats to their survival'
Who decides what is security and what is a threat?	Not necessarily only the state (though that is one major actor)	
3. To guide positive analysis	* JUT: 'Joined-up thinking' (Gasper, 2008)	Greatly expanded scope of analytical attention
4. To focus attention in policy design	Prioritising (which is inherent in the 'security' label): * *A focus on basic needs* * *Basic rights for all.* At the same time, Joined-up thinking → - broadly conceived policy response, and:	Policy style: (1) the large normative frame can influence other policy too (Werthes and Bosold, 2006: 23; promote coherence; (2) impossibility of unilateral control [their point c; 2006: 25] → 'Safety threats must be addressed through multilateral processes and by taking into account the patterns of

		interdependence that characterize the globalized world in which we are living' [point d; Bosold and Werthes, 2005: 89, 99]
Whose responsibility to respond?	Not necessarily only the state, which may lack the capacity	Werthes and Bosold (2006: 32): the focus on individuals appeals to a broader range of actors, not to states alone
5. To motivate	*Focus on basic needs and rights, including through an evocative label and concern for stability → stronger motivational basis, mobilizing attention and concern: sustaining Joined-up Feeling*	
What relation to discourses of development?	Goes further than discourses of human development, in the areas indicated above in *italics*	
What relation to discourses of need?	Relies on notion of need, as reasoned fundamental priority	
What relation to discourses of human rights?	*Basic rights for all*	

- awareness of impossibility of full knowledge of relevant factors → a deliberative, learning style in policy (Gasper and Truong, 2005)

Note: Italics indicate extensions beyond the Human Development Approach.

In the human security paradigm, a distinction may be drawn between *foundational* prevention and *crisis* prevention. [Ginkel and Newman, 2000] Foundational prevention is premised on the belief that prevention cannot begin early enough. It tries to address deep-seated causes of human insecurity. "Inequality, deprivation, social exclusion, and denial of access to political power are a recipe for a breakdown of social norms and order. Not having a fair chance in life ... being deprived of hope ... are the most incendiary root causes of violence and conflict". [Ginkel and Newman, 2000] To remove such causes requires a long-term strategy for equitable, culturally sensitive, and representative development. [Paragraph 51]

Preventive action is vastly more cost-effective than belated interventions to try to solve crises once they have exploded, for example, trying to supply emergency relief and build peace when a war has erupted (see Gasper, 1999). Lodgaard warned, however, that' 'textbook logic advocates preventive action while political logic suggests that action would have to wait till a crisis emerges' (paragraph 81); and that 'it is doubtful whether textbook logic and political logic can be reconciled unless the United Nations gets its own independent source(s) of finance' (paragraph 82).

In reaction to the record of political convenience and analytical ease being placed above policy coherence, the human security concept now serves "as a focal point around which an integrated approach to global governance is emerging" (Betts and Eagleton-Pierce, 2005: 7). Let us ask next: emerging from whom?

Roles for, and in relation to, whom? In motivation and mobilisation, the human security approach finds listeners more readily amongst some types of audiences than others. First, many general purpose international organisations, notably in the UN system, are seeking to integrate and make sense of their endeavours (and existence), and justify and prioritise their activities. This includes, not least, the UN system apex and UNESCO. In addition, some special purpose international organisations seek to identify key interconnections that decisively affect their area of responsibility and to identify the connections which show their own importance.

Second, some types of government have been attracted to the human security language: notably medium- and small-powers who are seeking a distinctive identity for their foreign policy, a purposefulness, meaningfulness and moral tone, and a niche for distinctive value addition. Since the human security framework draws attention to a great range of possible interconnections, it is perhaps not surprising that a relatively high proportion of observed users should be states, for their responsibilities span this range.

Third, we see uses by various social movements and civil society actors. The approach appeals to some progressive social movements trying to influence national and global policy directly or via influencing national and global society. It appeals to some feminists, and to a considerable variety of academics and intellectuals – in international relations, development studies, global social policy, public health, peace studies, etc. – seeking a policy-relevant intellectual framework for the twenty-first century.

Who has not adopted a human security language and framework? Relatively speaking, the big powers – compared to their degree of use of human rights discourse – but increased attention to global public goods might be changing this. Perhaps also private corporations, again in comparison to the degree of take-up of human rights language, but this too may be changing. In research circles, human rights studies does not seem much aware of its sister framework, while mainstream security studies has often resisted it, as we see later. Arguably, the framework has also been less taken up by national governments in their domestic analyses, compared to human development and human rights discourses. Lee stresses that "most Asian governments are unlikely to adopt a human security definition that contains political constraints or economic directives" (Lee, 2004: 37–8), i.e. that is seen to imply international rights to intervene or sanction a country in light of externally adjudged violations of either civil rights or economic-social rights, or to overrule countries' own cultures and traditions. The situation may be gradually changing. The very fear of undiluted human rights regimes makes some Asian governments prefer the more complex human security perspective. And while the national Human Development Reports that have taken human security as their theme are not directly owned by governments – the exercises have a quasi-autonomous status in order to ensure independent creative work – they have had significant government consultation and involvement.

Overall we could say that a human security perspective, like the thinking around human development, uses a global context and globally-oriented criteria of relevance. It tries to bring integration within the thinking of internationally-oriented agencies, by reference to priority criteria. In particular, it is guided by concerns with major threats and risks of crisis. According to Bosold and Werthes, the core use then of a human security approach has been in multilateral action to address priority threats to individual humans. Perception and formulation of what are the priority threats will vary. That flexibility provides space for diverse participants, and a sharper definition is not needed for a policy movement (Bosold and Werthes, 2005: 100–101).

Werthes and Bosold check how far the talk of the Human Security Network countries is only talk. They conclude that it has some real impact. It "has resulted in processes and developments which bring claims/pretension and substantiveness more in accordance with each other" (Werthes and Bosold, 2006: 28). As an example: after the success of the 1990s Ottawa process to ban anti-personnel landmines, the Network moved on to try to control trade in small arms and light weapons. This was done only with reference to illegal trade, for several leading members (Austria, Switzerland, Canada, even Norway; as well as observer South Africa) are major small arms exporters, and several have not been distinguished for their membership in or implementation of international agreements. Yet despite that restriction, the human security rhetoric is found to have provided a valuable instrument for holding its users accountable for their other actions (Werthes and Bosold, 2006).

Attacks on the idea of 'human security'

Attacks on the very idea: by claims about definition or about value priorities.

Some attacks on human security thinking concern the scope of issues it covers, but some object to the very notion, even when more narrowly conceived. Conventional security studies authors often assert that security is essentially a national level and military notion. Sometimes their claim is about established usage: 'human security emerged in a context in which security was predominantly conceived of in national terms", propose MacFarlane and Khong (2006: 233). In reality the term 'social security' is long and deeply entrenched, and the concept of psychological security has been in use for even longer (see Rothschild, 1995). MacFarlane and Khong themselves still adopt a notion of human security, though one of narrow scope, as we will see. Second, some claim that indisputable priority is a necessary condition for use of the 'security' label and that to use the term 'security' for non-military matters greatly overvalues their importance, which should be left to be judged instead in democratic elections. But then should not military threats also be judged through elections? Further, there is no reason why any prioritising mechanism will always prioritise military above non-military threats. The perspective of considering key threats to persons can be applied in many arenas. It is presumptuous for any one arena to claim proprietorial and exclusive rights.

A second form of attack proposes that security is a fearful ignoble craving, compared to the true ethical currency, freedom. The attack lacks foundation, for freedom rests on security, and, further, we wish to secure freedoms – though indeed basic freedoms, not everything. Both freedom and security have been emphases in elaborating human development discourse; both are prominent in for example Amartya Sen's work (e.g., Commission for Human Security, 2003).

Attacks on policy grounds: human security discourse is part of a dangerous agenda for world government, or no government – and is un-American

As with human rights discourse, mistrust comes from more than one side of the political spectrum. The G77 group of governments carry suspicions that human security discourse legitimates intervention by stronger powers. In contrast, a Heritage Foundation report on human security (Carafano and Smith, 2006) complains – despite having cited the commitment in the UN's founding Charter 'to employ international machinery for the promotion of the economic and social advancement of all peoples" – as follows:

Over the course of decades, the U.N. bureaucracy has come to see its role as facilitating not only peace and security, but also human rights, development, and social equity. . . . it is understandable that Americans question the U.N.'s seemingly constant pursuit of binding documents on themes that purportedly would advance security or development but in actuality would restrain U.S. power and leadership and undermine America's democratic and free-market practices. . . . Providing for the security and public safety of citizens is a principal attribute of national sovereignty. Indeed, nation-states that are democracies are best prepared to fill this role because their leaders are held accountable by the governed. . . . Shifting the focus of security policy from the collective will of free people to provide for their common defense to one of protecting a range of individual and collective political, economic, and cultural "rights" as defined by international bodies or non-state actors like NGOs confuses the nature of the modern state's roles and responsibilities.

(Carafano and Smith, 2006)

Similarly, MacFarlane and Khong (2006) insinuate that human security discourse can undermine the authority of the State, the only body able to do much about human security concerns. In reality human security discourse is clear on the primary role of the national State. The critics appear to believe implausibly that talk of any limits to the role of the State will undermine it.

Objections to a broad conception: further claims from definition

MacFarlane and Khong (2006) do not seek to restrict security language to the national level, but they attack the UN-UNDP-Japanese conception of human security which provides for inclusion of a broad range of threats. Sometimes, formally, they accept that allocation to threats of the priority status of 'security' language must depend on one's values, but in general they are not content with this.

First, they often presume terminological proprietorship. Thus environmental threats are explicitly marginalised: "the core of the debate on human security revolves around development and protection", they stipulate (p. 141). They try to reserve the term 'protection' exclusively for protection of life against violent attack, as if protection of health, and protection of anything else against anything else, does not constitute 'protection'. Proponents of such a narrow conception 'make the shift to the individual in theory, but ignore it in practice by subjectively limiting what does and does not count as a viable threat . . . [It] is communicable disease, which kills 18,000,000 people a year, not violence, which kills several hundred thousand, that is the real threat to individuals' (Owen, 2005: 38). Owen here means military style violence, and we should add that: 'It is estimated that each year 1.5 to 3 million girls and women are killed through gender-related violence' (Hamber *et al.*, 2006: 499).

Climatic movements combined with planned neglect by colonial regimes to leave tens of millions of people dead in the late 19th century (Davis, 2001); we face a parallel prospect in the 21st. MacFarlane and Khong's approach is thus better entitled a 'security studies approach' rather than 'protection-based'. It reflects the proprietorial claim that conventional 'security studies' feels toward the term 'security'.

Attacks on policy grounds: lack of prioritising power?

MacFarlane and Khong claim that broad human security discourse renders itself vacuous by including everything. Does it divert us from prioritisation? The work on Millennium Development Goals shows otherwise, both for prioritisation of areas and within areas. This operationalisation of parts of a human security perspective by Haq and his close associates (originally under the title 'International Development Targets' in the mid 1990s) centres on prioritising. MacFarlane and Khong fail to distinguish between prioritising between areas and within areas. Priority belongs not to a whole issue area *per se* but to basic levels of achievement therein. They recurrently misunderstand this, as in their attack on "redefining human development or health or environmental issues as security issues" (p. 264). Attainment and maintenance of the basic standards in these areas, but not of every matter in them, are issues of human security.

Prioritisation between areas is especially controversial. It represents the type of textbook logic that Lodgaard reminds us runs up against political 'logic', the convenience of established interests. For Picciotto *et al.* (2007) and Jolly *et al.* (2004) such comparisons are central. A broad-scope human security concept is needed to generate the required types of comparison: can we better promote security through military spending or through women's education or democracy education or . . .? Jolly reports for example how smallpox was eradicated in the late 1960s and 70s for just US $300 million, a cost equal to that of three fighter-bombers.[5]

While keen to prioritise between areas, human security analysis mistrusts invariable prioritisations of large areas. Beyond the elementary priorities, such as mass immunisation and access to oral rehydration therapy, it prefers a case-by-case approach. Broadness of general focus allows relevant prioritisation *in situ*, because one can then seek to identify the particular vulnerabilities that are actually prevalent, and felt as priorities, in particular cases (Jolly and BasuRay, 2007). Its broad approach is not a call for total analysis but for flexible analysis, instead of focusing by *a priori* disciplinary habit or prioritising by global over-generalisations.[6]

---

[5] Speech at the New School University, New York, 20 September 2007.

[6] For example, while a global econometric study might find no relation between inequality and conflict, in reality in some situations inequality may conduce to peace and in other situations to conflict, so that we need differentiation rather than a global generalisation. See also Barnett (chapter 3) on misdirection through over-generalised analysis.

Attacks on grounds of scope and explanatory force

MacFarlane and Khong (2006), Mack (2005) and others claim that the broad conception 'lacks analytical traction'. It adopts "the predictive/explanatory hypotheses that a broad set of aspects not conventionally connected in theory are often importantly connected in reality: including that the economic, social, cultural, medical, political and military are not separate systems; and that neither national nor personal security will be secured by military means alone" (Gasper, 2005: 228). A growing number of analysts, of many backgrounds, find this broader framing fruitful, though typically with some selective focusing according to the case considered. Health impact assessments of foreign policy, including on international economic relations, are one important illustration;[7] assessments of climate change's consequences for conflict are another.[8] The connections in Brauch's conceptual quartet (Figure 2.1) or any similar sketch mean that interest in any one member of the set will typically require deep attention to several of them.

Attacks on policy grounds: lack of influence?

In the short run, human security notions are often hard to apply in policy, because of problems concerning who cares and disagreements over who is responsible for action and who pays, reflecting the boundary crossing character of the issues considered. Mack (2005) proposes it is better to have a narrow vivid focus (on violent threats to individuals) because that captures attention and builds up sympathy which may later spread to dealing with other types of threat (MacArthur, 2007: 3); broad scope is considered not politically feasible in relation to rich country audiences. Implicit here is a short-run perspective that seeks immediate influence on current powerholders. Ignoring prevention and threats, other than physical violence, may in fact be shortsighted rather than hardheaded; it may lead not to eventual spread of concern, but to waste and later panic and evasion. Concentration on military interventions and subsequent 'patch-up/botch-up' efforts does not give a basis for building sympathy. It matches the short-run convenience of dominant interests in rich countries, who do not want to have causes of disasters traced far and fingers pointed at them.

In the short run, albeit perversely, the Japanese-backed broad picture such as in the Ogata-Sen Commission on Human Security was 'marginalized by the ongoing

---

[7] See a special issue of *Bulletin of the World Health Organisation*, March 2007, 85(3).

[8] Note, for example, the broadening of the range of threats and pathways considered in a 2007 CNA report on the security implications of climate change: 'The report includes several formal findings: Projected climate change poses a serious threat to America's national security; Climate change acts as a threat multiplier for instability in some of the most volatile regions of the world; Projected climate change will add to tensions even in stable regions of the world; Climate change, national security and energy dependence are a related set of global challenges.' (ECSP News, 14 June 2007, Woodrow Wilson Center; http://securityandclimate.cna.org/)

war on terror' (Bosold and Werthes, 2005: 97). The 'narrow' Canadian version appears to have been used at the 2005 World Summit of the United Nations, as well as by the UN Security Council in Resolution 1674 in April 2006 (MacArthur, 2007: 3). Responsibility to protect from severe threats of physical violence is taken on, but with no mention of other types of threat. The broader version so challenges vested interests that it represents a longer run agenda, just like human rights work has been since 1948.

Restrictiveness would endanger the human security perspective of interconnection, and is thereby less suitable as a perspective for research, mobilisation, and civil society engagement – the ways towards major long-run impact. Bosold and Werthes (2005) suggest that the narrow focus can be better for short-run campaigns on immediate easily graspable goals, like the land mines ban and the International Criminal Court; whereas the broader Japanese focus is better for the longer-run, since it sees deeper causes and effects, and can appeal to wider constituencies. As theorised in the Great Transition Initiative's scenarios of how a shift to more sustainable societies could eventually transpire (Raskin *et al.*, 2002), young people provide the energy for social movements, which generate and transmit the pressure and ideas for change, which can be picked up at times of eventual crisis and openness to reorientation when governments and other agencies must seek new responses. Discourses that make more radical points are likely to be ignored in short run policy, but have a different rationale and time perspective.

Concluding thoughts

The powerful opposition encountered by the broad human security discourse was our starting point, for why then does it continue to spread despite limited power-holder patronage? We looked at actual employment of the concept, since: "The meaning lies not in what people consciously think the concept means but in how they implicitly use it in some ways and not others" (Buzan *et al.*, 1998: 24). We have teased out a number of aspects in addition to the prioritising role of any 'security' concept:- the artificiality and arbitrariness of claims that security is exclusively a national level and military notion, and of attempts to restrict ideas of human security to one type of threat or one type of harm; the idea of basic normative threshold levels, across a range of needs, typically related to ideas of danger and vulnerability around causative threshold levels or tipping points in systems marked by ramifying interconnections; and the visceral charge of the idea of 'human security', as reflection of the vulnerabilities of human bodies, identities and personality.

We followed up the insight that 'human security' is a discourse, not merely a single concept. We highlighted an equity dimension, in which ideas from human needs, human development and human rights thinking are combined, including a priority to living a life of normal human span; and two connectivity dimensions, including a holistic perspective on real individuals' lives and a trans-disciplinary approach to explanation at the level of larger systems. Then we examined roles of this discourse: in generating situation-specific and unexpected insights, and in focusing policy design on foundational prevention rather than on palliative reaction to already erupted crises. It considered also who currently are the users and non-users of the approach.

The next section reviewed attacks on the idea of human security, especially on the broader versions. Against the claim that broad versions are unusable for analysis and explanation, we saw that they are increasingly used, typically combined with case-specific focusing, and can be dramatically insightful (see Picciotto *et al.*, 2007). Against the claim that broad versions are bad for establishing priorities, we saw that they emphasise prioritisation within sectors (as in the MDGs work) and, precisely thanks to their broad formulation, also between sectors. Against the claim that broad versions are politically impotent, we saw that while ramifying explanation tends to be unpopular with established interests, a short term orientation to immediate graspable goals is not the only relevant stance. A broader approach has potential for eventual broader and deeper support, towards longer term change.

Werthes and Debiel conclude that 'human security' is a political leitmotif. '[O]veremphasising the shortcomings of leitmotifs means to underestimate their potential, which exactly relies on its ambiguity/flexibility' (2006: 15; sic). This formulation is similar in spirit to Alkire's definition of the concept which was taken over by the Commission on Human Security. Thus, Japan can handle the leitmotif in a way that reflects its own history, culture and politics, with a focus on human needs and human development (Atanassova-Cornelis, 2006; Werthes and Debiel, 2006); whereas the EU must give a strong role to human rights in whatever human security orientation it adopts. Not every flexibly interpreted version of human security will attain impact in its environment. The Japanese and Canadian interpretations have led to some real movement, in different arenas, but whether the EU's human security talk makes any difference is still open to doubt (Werthes and Debiel, 2006: 18).

Werthes and Debiel helpfully point us to multiple users, interpretations, and uses. But their focus on direct policy uses by current policy users understates the potential of human security discourse, which has become a motivating framework in diverse sectors and professional contexts. Like some other commentators from international relations, they may insufficiently consider the 'human' perspectives in 'human security'. Human security thinking operates then both at more general levels – as a widely used concept, ideal and discourse in description, explanation and policy design – and at more concrete levels, as specified in particular research programmes and policy programmes. The more general levels of thinking inspire the more concrete and specific research and policy; they motivate integration across bound-aries: organisational, ideological and disciplinary. They do this in varied, unpre-dictable, niche-specific ways, as we see from the work in spheres such as violent conflict, AIDS and public health, climate change and migration. Concrete and precise research and policy programmes do not require that we establish a single narrow conception of human security, let alone one that is centred on safety from intentional physical violence. The causes and knock-on effects of damage through violence are so ramifying that while violence appears convenient as a focus for data collection and subsequent model-building, the associated research and policy are forced to ramify. A narrow frame provides no self-enclosed analytical coherence. We cannot afford to ignore wider causes and effects, and to treat the latter as externalities that will be absorbed by the human and natural environments. The world contains too much interconnection, fragility, and risk of straying past tipping points.

References

Alkire, S. 2003. A conceptual framework for human security. CRISE Working Paper 2, Queen Elizabeth House, University of Oxford.

Atanassova-Cornelis, E. 2006. Defining and implementing human security: the case of Japan. In T. Debiel and S. Werthes, eds., *Human Security on Foreign Policy Agendas: Changes, Concepts, Cases*. Duisburg: University of Duisburg-Essen, pp. 39–51.

Bauman, Z. 1994. *Alone Again: Ethics after Certainty*. London: Demos.

Betts, A. and Eagleton-Pierce, M. 2005. Editorial introduction: human security. *St. Antony's International Review*, **1**(2), 5–10.

Bilgin, P. 2003. Individual and societal dimensions of security. *International Studies Review*, **5**, 203–22.

Bosold, D. 2005/6. (Re-)Constructing Canada's Human Security Agenda. Available online: www.staff.uni-marburg.de/~bosold/pdf/Reconstructing_HumanSecurity_Oslo.pdf.

Bosold, D. and Werthes, S. 2005. Human security in practice: Canadian and Japanese experiences. *Internationale Politik und Gesellschaft/International Politics and Society*, **1**, 84–101.

Brauch, H. -G. 2005. Outline of *Globalisation and Environmental Challenges: Reconceptualising Security in the 21st Century*. Brauch, H. -G. *et al.*, eds., 2007. Berlin: Springer Verlag.

Burton, J. W. 1990. *Conflict: Basic Human Needs*. New York, NY: St. Martin's Press.

Buzan, B., Waever, O. and de Wilde, J. 1998. *Security: A New Framework for Analysis*. Boulder, CO: Lynne Rienner Publishers.

Carafano, J. J. and Smith, J. A. 2006. The Muddled Notion of 'Human Security' at the UN Heritage Foundation. Available online: http://www.heritage.org/Research/WorldwideFreedom/bg1966.cfm.

Collier, P., Elliott, V. L., Hegre, H. *et al.* 2003. *Breaking the Conflict Trap: Civil War and Development Policy*. Washington, DC: World Bank and Oxford University Press.

Commission on Human Security (CHS) 2003. *Human Security Now*. New York: UN Secretary-General's Commission on Human Security. Available online: http://www.humansecurity-chs.org/finalreport/.

Davis, M. 2001. *Late Victorian Holocausts*. London and New York: Verso.

Debiel, T. and Werthes, S., eds., *Human Security on Foreign Policy Agendas: Changes, Concepts, Cases*. Duisburg: University of Duisburg-Essen.

Edson, S. 2001. *Human Security: An Extended and Annotated Bibliography*. Cambridge, UK: Centre for History and Economics, King's College.

Gasper, D. 1999. Drawing a line: ethical and political strategies in complex emergency assistance. *European Journal of Development Research*, **11**(2), 87–115.

Gasper, D. 2005. Securing humanity: situating 'human security' as concept and discourse. *Journal of Human Development*, **6**(2), 221–45.

Gasper, D. 2007. Human rights, human needs, human development, human security: relationships between four international 'human' discourses. *Forum for Development Studies*, (NUPI, Oslo), 2007/**1**, 9–43.

Gasper, D. 2008. From 'Hume's Law' to policy analysis for human development: Sen after Dewey, Myrdal, Streeten, Stretton and Haq. *Review of Political Economy*, **20**(2), 233–56.

Gasper, D. and Truong, T -D. 2005. 'Deepening development ethics: from economism to human development to human security', *European Journal of Development Research*, **17**(3), 372–84.

GECHS 1999. *Science Plan: Global Environmental Change and Human Security*. Bonn: International Human Dimensions Programme (IHDP).

Government of Japan 1999. *Diplomatic Bluebook 1999*. Tokyo: Ministry of Foreign Affairs.

Hamber, B., Hillyard, P., Maguire, A. *et al.* 2006. 'Discourses in transition: re-imagining women's security'. *International Relations*, **20**(4): 487–502.

Haq, M. ul 1999. *Reflections on Human Development*, 2nd edition. Delhi: Oxford University Press.

Hough, P. 2005. 'Who's securing whom? The need for international relations to embrace human security. *St Antony's International Review*, **1**(2), 72–87.

Jolly, R., Emmerij, L., Ghai, D. and Lapeyre, F. 2004. *UN Contributions to Development Thinking and Practice*. Bloomington, IN: Indiana University Press.

Jolly, R., and BasuRay, D. 2007. Human security: national perspectives and global agendas', *Journal of International Development*, **19**(4), 457–72.

Lee, S. -W. 2004. *Promoting Human Security: Ethical, Normative and Educational Frameworks in East Asia*. Seoul: Korean National Commission for UNESCO.

Lodgaard, S. 2000. Human security: concept and operationalization'. Paper for UN University for Peace. Available online: http://www.upeace.org/documents/resources%5Creport_lodgaard.doc.

MacArthur, J. 2007. A responsibility to rethink? Challenging paradigms in human security. Paper to Symposium on Resolving Threats to Global Security, Dalhousie University, Halifax, NS, Canada.

MacFarlane, N. and Khong, Y. F. 2006. *Human Security and the UN: A Critical History*. Bloomington, IN: University of Indiana Press.

Mack, A. 2005. *Human Security Report*. Vancouver: University of British Columbia Press.

O'Brien, K. L., and Leichenko, R. M. 2007. Human security, vulnerability and sustainable adaptation. Background Paper for the UNDP Human Development Report 2007. New York, NY: UNDP.

Owen, T. 2005. Conspicuously absent? Why the Secretary General used human security in all but name. *St. Anthony's International Review*, **1**(2), 37–42.

Picciotto, R. 2005. Memorandum submitted to Select Committee on International Development. London: UK House of Commons. Available online: www.publications.parliament.uk/pa/cm200405/cmselect/cmintdev/464/5031502

Picciotto, R., Olonisakin, F. and Clarke, M. 2006. *Global Development and Human Security*. Stockholm: Ministry for Foreign Affairs, p. 14.

Picciotto, R., Olonisakin, F. and Clarke, M. 2007. *Global Development and Human Security*. New York, NY: Transaction Publishers/Springer, p. 24

Raskin, P., Banuri, T., Gallopin G. *et al.* 2002. *Great Transition*. Boston, MA: Stockholm Environment Institute.

Rothschild, E. 1995. What is security? *Daedalus*, **124**(3), 53–98.

Sen, A. K. 1993. Capability and well-being. In M. Nussbaum and A. Sen, eds., *The Quality of Life*. Oxford, UK: Clarendon, pp. 30–53.

UNDP 1990. *Human Development Report: 1990*. New York, NY: Oxford University Press.

UNDP 1993. *Human Development Report: 1993*. New York, NY: Oxford University Press.

UNDP 1994. *Human Development Report: 1994*. New York, NY: Oxford University Press.

UNDP 1996. *Human Development Report: 1996*. New York, NY: Oxford University Press.

UNDP 2002. *Bangladesh Human Security Report 2002: In Search of Justice and Dignity*. Dhaka: UNDP.

UNDP 2003. *Latvia Human Development Report: 2002–2003: Human Security*. Available online: http://www.undp.lv/?object_id=633.

Van Ginkel, H. and Newman, E. 2000. In quest of 'human security'. *Japan Review of International Affairs*, **14**(1), Spring 2000.

Weiss, T. G., Carayannis, T., Emmerij, L. and Jolly, R. 2005. *UN Voices: The Struggle for Development and Social Justice*. Bloomington, IN: Indiana University Press.

Werthes, S., and Bosold, D. 2006. Caught between pretension and substantiveness: ambiguities of human security as a political leitmotif. In T. Debiel and S. Werthes, eds., *Human Security on Foreign Policy Agendas: Changes, Concepts, Cases*. Duisburg: University of Duisburg-Essen, pp. 21–38.

Werthes, S. and Debiel, T. 2006. 'Human security on foreign policy agendas. introduction. In T. Debiel and S. Werthes, eds., *Human Security on Foreign Policy Agendas: Changes, Concepts, Cases*. Duisburg: University of Duisburg-Essen, pp. 7–20.

Wolfers, A. 1962. National security as an ambiguous symbol. In A. Wolfers, eds., *Discord and Collaboration*. Baltimore, MD: John Hopkins University Press, pp. 147–65.

3

Climate change science and policy,
as if people mattered

JON BARNETT

Introduction

There is something troubling about the bulk of climate change research and policy, particularly when viewed from the position of the Pacific Islands. A number of authors have attempted to articulate their disquiet about the mainstream of climate change impacts research, arguing that it has tended to reduce societies to simple and predictable systems able to be comprehended through large-scale aggregating models, and able to be managed through technical and rational strategies (see Taylor and Buttel, 1992; Proctor, 1998; Rayner and Malone, 1998; Shackley et al., 1998; Agrawala, 2001; Demeritt, 2001; O'Brien et al., 2004; Thrift, 2004). The result is that climate change has been portrayed as an environmental problem with somewhat separable human dimensions, suggesting in turn adaptation actions that favour environmental and technical rather than social and institutional changes. Further, for all the effort that has gone into it, the reduction in emissions that will be achieved by the Kyoto Protocol will do almost nothing to slow the rate of climate change, and it falls far short of the 75% reduction below current levels of emissions that is necessary to avoid 'dangerous' climate change in the Pacific Islands and elsewhere. The new round of 'global horse trading' that has begun over future emissions targets does not inspire confidence that the necessary cuts in emissions will be forthcoming (Najam et al., 2003: 224). The slow progress on the implementation of the Adaptation Fund and the voluntary nature of the Least Developed Countries Fund and the Special Climate Change Fund suggests that adaptation will be poorly funded in the future.

It is these shortcomings in research and more so in policy that have led to new suggestions for understanding and acting on climate change. There is a growing consensus on the need to move away from top-down model-driven studies of large areas to bottom-up participatory approaches in smaller places. These 'second generation' (Burton et al., 2002) studies can explain in detail the ways and extent to which climate change is a human security problem, and can identify appropriate and effective adaptation options. In policy terms, common themes of new proposals include: (1) to engage broader policy domains such as finance ministries and development agencies; (2) to base the allocation of future targets more firmly on issues of justice and equity than political expediency; (3) to engage with a deeper range of constituencies beyond national governments; and (4) to begin implementing adaptation actions (see Aldy *et al.*, 2003; Najam *et al.*, 2003; Dessai *et al.*, 2004; Yamin, 2004; Christoff, 2006; Paavola *et al.*, 2006).

These new kinds of approaches to research and policy are not evident in (or when viewed from) the South Pacific region. This is of concern, given that the countries in the region are consistently identified as among the most vulnerable of all countries to climate change. This chapter explains some of the reasons why this is the case, principally by critically examining some of the main trends in research on the impacts of climate change. It argues that the dominance of modelling approaches to climate impacts in the region is now a barrier to the recognition of and development of solutions to climate insecurity in the region. It proposes instead a human security approach to climate change research and policy in the region, and it briefly describes what this might entail.

Understanding climate change: big pictures and local lacunas

There is now no doubt that the earth's climate is warming, and that this is 'very likely due to the observed increase in anthropogenic greenhouse gas concentrations' (IPCC, 2007a: 10). The changes now underway have no precedent in the history of civilisation. Anticipated changes include increasing mean temperatures, rising average sea levels, more frequent hot spells, more heavy precipitation events, an increase in the area affected by drought, more intense cyclones and increasing incidences of high sea level events (IPCC, 2007a). There is some uncertainty about the magnitude of these changes, for example, the range of possible temperature increases is between 1.1 and 6.4°C by the year 2100, and the range of possible increases in sea level is between 18 and 59 cm by the year 2100 (although this excludes the possibility of melting from ice sheets, the consequences of which could be sea level rise in excess of 1 m by the end of the century) (IPCC, 2007a).

These changes in climate will affect ecological systems in key ways. Already there are confident observations of glacial lakes growing in size and number, and increases in spring run-off from glacier and snow-fed rivers; increasing instability of ground areas in permafrost regions (implying melting); changes in Arctic and Antarctic ecosystems; warming of lakes and rivers; earlier timing of spring events across many ecosystems; shifts in a number of plant and animal species towards the poles; and shifts in ranges and abundance of marine organisms, including plankton and fish (IPCC, 2007b). These changes are likely to be due to climate change, and they are likely to increase in the future as the climate continues to warm. The social consequences of these changes have been talked about as security problems. In these discussions 'security' has remained a rather indistinct concept, relating largely if not explicitly to the way climate change may cause economic damages within countries rather than to the ways in which it might be a factor in armed conflict between countries (Barnett, 2003).

There has of course long been some suggestion that climate change might be a cause of war (see Homer-Dixon, 1991; Gleick, 1992; van Ireland *et al.*, 1996), but it is only recently that the prospect of climate-induced conflicts has been popularised, particularly through a study commissioned by the US Department of Defense (Schwartz and Randall, 2003) and a later study by 11 retired admirals and generals (CNA Corporation, 2007). These neo-Malthusian visions are very familiar to students of critical geopolitics and environmental security. They differ from classical geopolitics only in as much as the latter saw the 'stage' of geography as the fixed parameters that shape the social 'play', whereas the climate geopolitical scripts are predicated on a claim to knowing how changes in the stage will drive changes in the social play. Their principal discursive function is to sustain the legitimacy of armed forces in an age where the sources of insecurity are much less obviously weapons (Barnett, 2001; Dalby, 2002). For example, the CNA (2007: 7) study recommended, among other things, that the security implications of climate change should be 'fully integrated into national security and national defence strategies' and that 'the Department of Defense should enhance its operational capability ... that result[s] in improved US combat power through energy efficiency' (CNA, 2007: 8). That the main advocates of this view are institutions within the US foreign policy community suggests that 'climate security' is more likely to be understood as a conventional national security problem rather than as a human security problem (as has been the case with 'environmental security' more generally [Barnett, 2001]).

It is well understood that resource-dependent and low-income societies are typically highly at risk from climate change (see Bohle *et al.*, 1994; Adger, 1999; Kates, 2000; Leichenko and O'Brien, 2008). Yet, the future impacts of climate change on specific social systems are uncertain, and this is a function of uncertainty about the magnitude of changes in ecosystems upon which people depend and uncertainties about the capacity of social systems to adapt to these changes. Nevertheless, the kinds of ecological changes that are likely justify considering climate change as a grave threat to human security. For example, the 2007 report from the Intergovernmental Panel on Climate Change (IPCC) describes a wide range of likely climate impacts that will undermine human security, including an increase in drought-affected areas, affecting, for example, up to 250 million people in Africa; decreasing flows in rivers that supply water to millions in Latin

America and a billion people in Asia; declining crop productivity in low latitudes, including a 50% decline in yields in some parts of Africa and 30% decline in yields in some parts of Central and South Asia; millions of people exposed to flooding in the densely populated and economically productive mega deltas of Asia; increases in malnutrition in low-income societies; increased deaths, diseases and injuries associated with extreme events, such as droughts, floods, heatwaves, fires and storms; decreasing yields of fish from most of the world's freshwater and coastal fisheries; and loss of lands, homes and possibly islands in many of the small island states in the South Pacific, Caribbean and Indian and Atlantic oceans (IPCC, 2007b).

In some cases there are limits to adaptation to climate change, which reinforce the idea that catastrophic social losses are possible. For example, if recent estimates of a 140 cm rise in sea-level rise (Rahmstorf, 2007) and annual coral bleaching (Donner *et al.*, 2005) are correct, then there is little that can be done to avoid or adapt to losses of land on low-lying atoll islands. The result may be increases in morbidity and mortality, and increased demand for migration, with a worst case outcome being the collapse of the ability of island ecosystems to sustain human habitation and sub-sequent risks to the sovereignty of the world's five atoll-island states (Barnett and Adger, 2003). In the Arctic, too, there is arguably little that can be done to avoid or adapt to absolute losses of snow and ice, and the changes in social–ecological systems, including increased morbidity and mortality and migration, that may result. In both atolls and the Arctic there are other significant losses as well, including of place and culture and the right to a nationality and a home (Adger *et al.*, 2009). In each case migration cannot be seen as an 'adaptation' but rather as a loss – of culture, livelihood, place and the right to a home.

So, many social systems are thought to be vulnerable to climate change for reasons of dependence on climate sensitive ecosystems and a lack of the entitle-ments thought necessary to adapt to changes in these systems, and there is some general sense of the adaptations that may be required to avoid these impacts. Yet there remains a lack of detailed empirical research into the nature of vulnerability and adaptation in specific places and on specific social groups. To be sure, there is a considerable body of useful work that outlines conceptual and theoretical frame-works that help explain why climate change is a social problem (see Burton, 1997; Handmer *et al.*, 1999; Smit *et al.*, 1999, 2000; Smith *et al.*, 2001). There are also socially oriented national- and regional- scale studies that point to the risks that climate change poses to human security (see Woodward *et al.*, 1998; Bryant *et al.*, 2000; Leichenko and O'Brien, 2002; O'Brien *et al.*, 2004; Barnett *et al.*, 2007). However, compared to modelling-based approaches to assessing impacts and adap-tation, there are few studies that are informed by detailed investigations of the lives and livelihoods of specific groups of people in particular locations (see Adger,

1999; Eakin, 2000; Ford *et al.*, 2006; Leary, *et al.*, 2006; Tschakert, 2007; Tyler *et al.*, 2007).

The Pacific lacunae

If this last statement is contentious when applied to climate impacts research in general – and it will increasingly be, given the rapid proliferation of such studies in the pages of journals such as *Global Environmental Change* – it is not contentious when applied to the Pacific Islands. Chapter 17 of the Intergovernmental Panel on Climate Change's 2001 assessment report worth quoting at length. On this subject:

Finally, there is some uneasiness in the small island states about perceived over reliance on the use of outputs from climate models as a basis for planning risk reduction and adaptation to climate change. There is a perception that insufficient resources are being allocated to relevant empirical research and observation in small islands. Climate models are simplifications of very complex natural systems; they are severely limited in their ability to project changes at small spatial scales, although they are becoming increasingly reliable for identifying general trends. In the face of these concerns, therefore, it would seem that the needs of small island states can best be accommodated by a balanced approach that combines the outputs of downscaled models with analyses from empirical research and observation undertaken in these countries.

(Nurse and Sem, 2001: 870)

In the period since this report was published, there has been very little activity to address this problem in the South Pacific region. There have since been very few critical academic publications on social vulnerability and adaptation to climate change in the region, and indeed there has been very little new research on any aspect of climate change in the region apart from some new studies on coral reefs, which builds on some highly valuable earlier field-based studies of coastal vulnerability. There have been two major regional projects in recent years: the successful and socially oriented CDN $2.2 million Capacity Building for the Development of Adaptation in Pacific Islands Countries (CBDAMPICS) project funded by the Canadian International Development Agency, which conducted studies in four countries (see Sutherland *et al.*, 2005); and the model-driven Pacific regional project under the larger Assessments of Impacts and Adaptations to Climate Change (AIACC) project. There have also been 'second generation-style' projects associated with the National Adaptation Plans of Action being prepared in the region's Least Developed Countries of Kiribati, Samoa, the Solomon Islands, Tuvalu and Vanuatu, although some of these, like the AIACC project, have experienced difficulties of various kinds. Given that the South Pacific region contains 24 countries and territories and 20% of the world's languages, and its islands have long been recognised as among the most vulnerable of all places to climate change, that almost nothing is being done to understand the likely social impacts of climate change and the possible responses to it is a moral and intellectual problem of considerable importance.

It is tempting to suggest that the reason why more locally oriented empirical studies of climate impacts are not being conducted in the Pacific is because studying something that is only beginning to happen is challenging. However, this is not really the case: there are observations of change in climate and ecosystems (see Salinger *et al.*, 2001; MNREM, 2005; MELAD, 2007); there are clear and practical methodologies now available for conducting such assessments (see Lim and Spanger-Siegfried, 2004); and there are many lessons to be learned from studying existing sensitivities to current climate (Glanz, 1988). The reasons cannot be because of cost either, since local-level studies are typically no more expensive than modelling studies, for while they are often more time consuming, they are far less capital and technology intensive. Nor could it be said that the Pacific Islands are little studied because they are of marginal interest in climate change terms (as they are in many other global problems, such as the 'war on terror' or the 'war on drugs'). Indeed, they are a *cause célèbre* of the international climate research and NGO communities; barely a month goes by that the issue of 'environmental refugees' from the Pacific and highly vulnerable countries like Tuvalu is not in the media of one or more OECD countries (Farbotko, 2005). Indeed, there is a nascent disaster tourism in the Pacific because of climate change, and in many countries climate change officers spend much of their time dealing with (or avoiding) the international media.

Rather, the problem is to do with the hegemony of modelling approaches to impacts and the control of impacts research by advocates of models. Of particular concern is the use of Integrated Assessment Models (IAMs) at almost every opportunity to conduct research or training on climate impacts research in the region. In the Pacific, IAMs and their advocates are ubiquitous, they appear when-ever there is funding for impacts research, and they almost always prevail because they claim the mantle of 'science'. Their advocates typically come from or have been trained in developed countries and are far more entrepreneurial than Pacific Islanders in negotiating research funding. IAMs seek to integrate information through mathematical representations of aspects of natural and social systems (Risbey *et al.*, 1996), and those such as VANDACLIM and its Cook Islands, Fijian and Kiribati versions (see Warrick *et al.*, 1999), and PACCLIM (see Kenny and Ogoshi, 1999) were heavily promoted and widely used throughout the first major regional climate change project and the preparation of initial National Communications to the UNFCCC.

That the hegemony of modelling has been a problem is alluded to in the above quote from Nurse and Sem (2001), and while the modelling agenda is now privately viewed with more scepticism by most island-based agencies, it seems that little else has emerged from regional research agencies and global agencies tasked with assisting the region. The AIACC project for the region, for example, describes itself as seeking 'to develop the "next generation" of integrated assessment methods and models, for application at island and sub-island scales' and these include as one of many features 'human dimensions components'.[1] Its almost complete failure to produce has been seen to be due to the 'unavailability or the dubious quality of some data' (AIACC, 2003). In June 2007, the IPCC Working Group I Task Group on Data and Scenario Support for Impact and Climate Analysis held a 3-day Expert Meeting on Regional Impacts, Adaptation, Vulnerability and Mitigation. The call for papers identified six key areas including 'integrating data sets', 'exploring feedbacks and couplings among different systems' and 'identifying spatial teleconnections'.[2] Not only is this a language all but a few regional researchers can understand, the flow of information at the meeting was unidirectional, because, as the call for papers explained, the 'leading experts ... will present keynote talks ... with ample time included for discussion', whereas authors of accepted abstracts will be allowed 3 minutes in which to present a poster.

Thus the model of science in the region remains the same: 'experts' and their top-down generic models set the norms for knowledge and its production. In doing so they marginalise the value of endogenous knowledge and approaches to knowledge generation, and crowd-out the intellectual and resource space for local approaches and knowledge to emerge. When data and/or technology is not available, when the vernacular of modelling science is not comprehendible, and when the utility of generic models is questioned, the deficiencies are always seen as being in and of the islands, since the standards of climate science are presumed to be universal and its modes of expansion and delivery are rarely questioned.

Yet models, in particular Integrated Assessment Models, are far from ideal tools for studying climate impacts on islands. They were used in the region – and at considerable expense – as training and information tools for the preparation of initial National Communications to the UNFCCC. However, while they yield useful results about potential climate impacts at large scales and sectors, the results are of less use to the countries themselves, and it is striking the degree to which models *did not* inform the national communications of most countries. Countries and researchers have since noted the limitations of modelling tools, including that they were too generic and not flexible enough to accommodate the diverse social

---

[1] http://www.aiaccproject.org/aiacc_studies/aiacc_studies.html
[2] http://ipcc-wg1.ucar.edu/meeting/TGICA-Regional/TGICA-Rgnl_public.html

and ecological characteristics of countries; consumed too much time and money; required external 'expert' knowledge and tended to ignore or discount local knowledge; assessed units at the scale of one or more grid squares in General Circulation Models that are between 200 and 600 km^2, which provides insufficient resolution for the land areas of small islands; were overly focused on future climate impacts and tended to ignore or downplay current vulnerabilities; were so focused on uncertainties that they downplayed or ignored those things that are relatively certain; did not give enough recognition to the problem of climatic extremes; tended to consider all adaptations as being technically possible, without considering the practical constraints to their implementation; and ignored the needs and voices of local communities (based on author's observations, as well as Kaluwin and Smith, 1997; Barnett, 2002; Burton *et al.*, 2002; Lim and Spanger-Siegfried, 2004).

Perhaps the most important problem with models is that they are an expression of an extremely modern cosmology that is incompatible with Pacific cosmologies. In the Pacific nature is indivisible from the social; the economic, cultural and political are indivisible from genealogy; and the world is less a series of contracting spaces and more a set of expanding opportunities (see Hau'ofa, 1994; Kempf, 1999; Banks, 2002). Models homogenise places and social groups, treat them as spatially and geographically bounded, see social life as the sum of rational individual actions, assume 'culture' is separable from other aspects of society; and assume that nature and society are independent 'facts' (Morgan and Dowlatabadi, 1996; Proctor, 1998; Shackley and Gough, 2002). Models therefore tend to produce a knowledge of the Pacific Islands that is alien to and alienates Pacific Islanders.

There are therefore some lacunae in climate impacts research – in the Pacific, if not elsewhere – that stem from and are sustained by the power of modelling science that comes from developed countries. Addressing these lacunae through alternative methodologies for researching climate impacts that reflect local concerns and contexts, is therefore necessary to more powerfully communicate to the international community the losses that may arise in islands if emissions of greenhouse gases are not reduced; to contextualise and communicate vulnerability in the particular contexts in which it arises; to determine effective and legitimate adaptation strategies that are locally suitable; and to empower Pacific Islanders to 'own' climate change. In short, some decolonisation of climate impacts research is required.

New approaches for assessing vulnerability and adaptation to climate change have emerged. These have been called 'second generation', 'vulnerability/adaptation' or 'bottom-up' approaches. They include the United Nations Development Programme's *Adaptation Policy Framework* (Lim and Spanger-Siegfried, 2004), the *National Adaptation Plan of Action Guidelines* (UNFCCC, 2002) and the method used in the aforementioned CBDAMPIC project. These new approaches recognise the diversity in the social and environmental conditions of countries. They therefore do not prescribe 'a common methodology', but rather offer a framework of linked concepts, key questions, methods and principles for assessment than can be combined in various ways to suit the conditions of any given country. They differ in key ways from earlier methodologies in that they focus their analysis on current vulnerability to present day climate; focus on smaller scales of social organisation where decisions about adaptation are and will be made; prioritise social systems by focusing on the present and future social and economic forces that create vulner- ability; are concerned with delivering grounded assessments of adaptation actions; emphasise the problem of climate extremes (as opposed to changes in mean condi- tions); include stakeholders in assessments of vulnerability and adaptation; integrate a wider range of existing studies and information on, for example, resource management, planning, economic development, household expenditure and decision-making processes; and consider the capacity of social systems – including the policy process – to implement adaptation actions.

This 'bottom-up' approach is one that is better suited to Pacific Island Countries. It means that the information that is required is information that people already possess, traditional knowledge is valued, that the capacity to conduct assessments lies within countries, and countries can have more confidence in and ownership of the results of assessments and their proposed adaptation actions. Their emphasis on the local context, the social and economic forces that affect places, and the strategies to manage existing climate extremes, makes them highly compatible with approaches to assessing human (in)security.

Human security: position, power and perspective

The concept of human security unites international relations and development theory and practice. From the international relations side, the end of the Cold War, advances in communication technologies, increasing economic interdepend- ence and environmental change, among other factors, have challenged the state's monopoly over the meaning and practice of 'security'. Indeed, human security highlights the possibility that what states do in the name of 'national security' endangers their own and other people in important ways. As development theory and policy uses the term, human security is also about the ways myriad processes, such as unemployment, trade deficits and changes in food prices, can undermine people's security. In tracing through these pathways, human security analyses demonstrate that some people's security occurs at the expense of others (Booth, 1991). As security has become more pluralised in this way – away from states and away from war, and towards people and the multitudinous risks they must manage – it increasingly becomes a general concept of social science (Shaw, 1993).

Human security synthesises concerns in development theory and practice for basic needs, human development and human rights (Gasper, 2005). The concept came to prominence through the 1994 *Human Development Report*, which defined human security as a 'concern with human life and dignity' (UNDP 1994: 22), and which adopted a comprehensive approach by identifying economic, food, health, environmental, personal, community and political components to human security. The orientation is therefore firmly on human beings, and, in this early formulation, on basic needs ('human life') as well as psychosocial elements of being ('dignity'). Through the use of the word 'security', this and later formulations of human security point to the need for the things that are important to human life and dignity to be maintained, despite sudden and incremental changes in the social and environmental milieu that determine their provision.

The International Commission on Human Security defines human security as being about protecting 'the vital core of all human lives in ways that enhance human freedoms and human fulfilment' (Commission on Human Security, 2003: 4). This definition continues the focus on human dignity ('fulfilment') and builds on Amartya Sen's (1999) groundbreaking work on the importance of freedoms to human development. Sen argues that development is not so much something that can be done to others, but is instead something that people do *for themselves* given sufficient 'economic opportunities, political liberties, social powers, and the enabling conditions of good health, basic education, and the encouragement and cultivation of initiatives' (1999: 4). These opportunities are, in Sen's words, 'freedoms', and it is freedom, he argues, that should be both the means (how to attain) as well as the ends (the goal) of development.

The idea that there is a '*vital* core' whose degradation is a *security* problem points to the way in which human security is different to human development, in the same way that an 'environmental security' problem differs from an environmental problem. The concept of 'security', regardless of referent object or risk, entails differentiation of 'security' issues from everyday 'low politics' issues. This process is known as 'securitisation' and it is a 'speech act' that raises the status of an issue from ordinary to extraordinary (Waever, 1995). The identification of the critical problems that warrant the label 'security' is something that is best left to the groups that must contend with them since different groups have different values that lead to different prioritisations of problems. In their approaches Sen and the Commission on Human Security avoid the problem of value homogenisation that arises from detailed prescriptions of what is good for people and communities. But this is not to say that they eschew the idea of basic needs and fundamental rights. There are indeed basic needs, such as access to nutritious food and clean drinking water, and basic rights, such as the freedom from personal injury and forced migration, that are essential to every life, and there is nothing in either Sen's or the Commission's work that denies this. Nor does recognition of value pluralisation suggest that violation of these needs and rights is not a human security problem, and indeed they should be considered as such until demonstrated otherwise. But it does suggest that at some point beyond basic needs and rights identification of human security requires some prioritisation of issues and that this is best done by groups that must manages these issues.

A human security approach to climate change would entail focusing on the effect of climate change on the well-being of people and communities, which may be critically influenced by mediating institutions such as the markets and the state, but which cannot be understood merely by analysing the effects of climate change on these larger categories. It would also involve analytical integration of multiple drivers of human security, an insistence on basic human needs, rights and responsibilities, and a concern for justice (Gasper, 2005). Importantly for climate change research, where 'vulnerability' refers to losses of a generic kind and whose relative significance is rarely considered, and where the purpose of adaptation actions is rarely specified, a human security approach to climate change would ask 'what matters most here' as a means to assess what potential losses may matter most and what adaptation actions should be prioritised. This need to consider values and prioritisation of issues to identify the difference between 'security' problems and 'low politics' problems is a major point of difference between a human security approach to climate change and other approaches. It may be eschewed by climate change researchers since it is the case, largely because of the way poverty tends to lead to discounting of the future, that many people in developing countries do not prioritise climate change ahead of more basic problems such as access to clean water, lack of access to education, and personal safety. This is not to say that the risks climate change poses to values cannot be meaningfully identified, but it is to say that simply asking poor people if climate change is a problem tends not to produce the answer climate change researchers might hope to hear.

Conclusions

Research on climate change in the Pacific Islands has failed to adequately demonstrate the ways in which it is a human security issue, even though these are places that are often cited as being among the most at risk from climate change. A barrier to doing the kinds of bottom-up research that can identify climate change as a human security problem is the prevalence of modelling approaches, whose dominance precludes more site-specific and socially oriented research from taking place. This might not be so much of a problem if the models that have been applied in the region were able to deliver policy-relevant insights into vulnerability and adaptation; but they have not, as measured by both the lack of peer-reviewed publications of their results and their lack of influence in the reports and policies prepared by most Pacific Island Countries. Indeed, for the most part, IAMs work against the recognition and the development of solutions to climate insecurity in the Pacific Islands.

That climate change is not seen as a human security problem for Pacific people has implications for policy too. One value of a human security approach to climate change is the way it can *humanise* risks to decision makers. Based on available research – largely about the vulnerability of coastal systems – the Small Island States have argued for large cuts in emissions. However, there is very little research on social vulnerability that they could use to construct more powerful arguments about potential social losses, which might take the form of relating potential impacts to the international human rights instruments or pointing to the risks climate change poses to development programmes. Further, there is very little systematic research that governments can use to determine and prioritise adaptation options. The absence of this specific information about social vulnerability and adaptation means that there is insufficient recognition of the real magnitude of climate dangers in the Pacific and less impetus for emissions reductions, and it enables OECD countries to argue for delaying assistance for adaptation on the grounds that the information about adaptation possibilities is insufficient. Of course, this problem of a lack of information about climate change as a human security problem in the South Pacific is only one of many reasons for the tardiness of the international community to reduce emissions to avoid dangerous climate change and to assist the Pacific Islands to adapt, but it is an important one given that what little power these countries have in climate politics comes largely from the moral pressure they are able to exert (Paterson, 1996, Shibuya, 1996). A human security approach to climate change can reinforce these moral arguments and can better inform policy and planning for adaptation.

References

Adger, W. N. 1999. Social vulnerability to climate change and extremes in coastal Vietnam. *World Development*, **27**, 249–69.

Adger, N., Barnett, J. and Ellemor, H. 2009. Unique and Valued Places at Risk. In Schneider, S., Rosencranz, A. and Mastandrea, M, eds., *Climate Change Science and Policy*. Washington, DC: Island Press, pp. 131–138.

Agrawala, S. 2001. Integration of human dimensions in climate change assessments. Presentation at Open Meeting of the International Human Dimensions of Global Change Community. Rio de Janeiro, 6–8 October 2001.

AIACC 2003. *AIACC Project SIS09 Progress Report for July–December 2003*. Available online: http://sedac.ciesin.columbia.edu/aiacc/progress/SIS09_July03.pdf.

Aldy, J., Ashton, J., Baron, R. *et al.* 2003. *Beyond Kyoto: Advancing the International Effort Against Climate Change*. Arlington, VA: Pew Centre on Global Climate Change.

Banks, G. 2002. Mining and the environment in Melanesia. *The Contemporary Pacific*, **14** (1), 39–67.

Barnett, J. 2001. *The Meaning of Environmental Security*. London: Zed Books.

Barnett, J. 2002. The challenges of adapting to climate change in Oceania. *Pacific Ecologist*, **1**, 25–8.

Barnett, J. 2003. Security and Climate Change. *Global Environmental Change*, **13**(1), 7–17.

Barnett, J. and Adger, W. N. 2003. Climate dangers and atoll countries. *Climatic Change*, **61**, 321–37.

Barnett, J., Dessai, S. and Jones, R. 2007. Vulnerability to climate variability and change in East Timor', *Ambio*, **36**(5), 372–8.

Bohle, H., Downing T. and Watts, M. 1994. Climate change and social vulnerability: toward a sociology and geography of food insecurity. *Global Environmental Change*, **4**, 37–48.

Booth, K. 1991. Security and emancipation. *Review of International Studies*, **17**(4), 313–26.

Bryant, C., Smit, B., Brklacich, M. *et al.* 2000. Adaptation in Canadian agriculture to climatic variability and change. *Climatic Change*, **45**(1), 181–201.

Burton, I. 1997. Vulnerability and adaptive response in the context of climate and climate change. *Climatic Change*, **36**, 185–96.

Burton, I., Huq, S., Lim, B., Pilifosova, O. and Schipper, L. 2002. From impacts assessment to adaptation priorities: the shaping of adaptation policy. *Climate Policy*, **2**, 145–59.

Christoff, P. 2006. Post-Kyoto? Post-Bush? Towards an effective 'climate coalition of the willing'. *International Affairs*, **82**(5), 831–60.

CNA Corporation 2007. *National Security and the Threat of Climate Change*. Alexandria, VA: CNA Corporation.

Commission on Human Security 2003. *Human Security Now*. New York, NY: Commission on Human Security.

Dalby, S. 2002. *Environmental Security*. Minneapolis, MN: University of Minnesota Press.

Demeritt, D. 2001. The construction of global warming and the politics of science. *Annals of the Association of American Geographers*, **91**, 307–37.

Dessai, S., Adger, N., Hulme, M. *et al.* 2004. Defining and Experiencing Dangerous Climate Change. *Climatic Change*, **64**(1–2), 11–25.

Donner, S., Skirving, W., Little, C., Oppenheimer, M. and Hoegh-Guldberg, O. 2005. Global assessment of coral bleaching and required rates of adaptation under climate change. *Global Change Biology*, **11**(12), 2251–65.

Eakin, H. 2000. Smallholder maize production and climatic risk: a case study from Mexico. *Climatic Change*, **45**, 19–36.

Farbotko, C. 2005. Tuvalu and climate change: constructions of environmental displacement in the Sydney Morning Herald. *Geografiska Annaler*, **87**(4), 279–93.

Ford, J., Smit, B. and Wandel, J. 2006. Vulnerability to climate change in the Arctic: a case study from Arctic Bay, Canada. *Global Environmental Change*, **16**(2), 145–60.

Gasper, D. 2005. Securing humanity: situating 'human security' as concept and discourse. *Journal of Human Development*, **6**(2), 221–45.

Glanz, M. 1988. *Societal Responses to Regional Climate Change: Forecasting by Analogy*. Boulder, CO: Westview Press.

Gleick, P. 1992. Effects of climate change on shared fresh water resources. In I. Mintzer, ed., *Confronting Climate Change*. Cambridge, UK: Cambridge University Press, pp. 127–40.

Handmer, J., Dovers, S. and Downing, T. 1999. Societal vulnerability to climate change and variability. *Mitigation and Adaptation Strategies for Global Change*, **4**(3/4), 267–81.

Hau'ofa, E. 1994. Our Sea of Islands. *The Contemporary Pacific*, **6**, 147–61.

Homer-Dixon, T. 1991. On the threshold: environmental changes as causes of acute conflict. *International Security*, **16**, 76–116.

IPCC (Intergovernmental Panel on Climate Change) 2007a. *Climate Change 2007. The Physical Science Basis: Summary for Policymakers*. Geneva: IPCC Secretariat.

IPCC (Intergovernmental Panel on Climate Change) 2007b. *Climate Change 2007. Climate Change Impacts, Adaptation and Vulnerability*. Geneva: IPCC Secretariat.

Kaluwin, C. and Smith, A. 1997. Coastal vulnerability and integrated coastal zone management in the Pacific Islands Region. *Coastal Research*, **24**, 95–106.

Kates, R. 2000. 'Cautionary tales: Adaptation and the global poor. *Climatic Change*, **45**(1), 5–17.

Kempf, W. 1999. Cosmologies, cities, and cultural constructions of space: oceanic enlargements of the world. *Pacific Studies*, **22**(2), 97–114.

Kenny, G. J. and Ogoshi, R. 1999. Climate change and Pacific island agriculture: impacts, vulnerability and adaptation. In: International Global Change Institute (IGCI) (1999). PACCLIM Workshop Proceedings: Climate Change and Sea Level Rise in the Pacific: Impacts and Adaptation. Proceedings of Workshop held in Auckland, 23–27 August 1999. Sponsored by Asia-Pacific Network and The World Bank.

Leary, N., Adejuwon, J., Bailey, W. *et al.* 2006. *For Whom the Bell Tolls: Vulnerability in a Changing Climate. A Synthesis from the AIACC Project*. AIACC Working Paper No. 21. Florida: International START Secretariat.

Leichenko, R. and O'Brien, K. 2002. The dynamics of rural vulnerability to global change: the case of southern Africa. *Mitigation and Adaptation Strategies for Global Change*, **7**(1), 1–18.

Leichencho, R. M. and O'Brien, K. L. 2008. *Environmental Change and Globalization: Double Exposures*. New York, NY: Oxford University Press.

Lim, B. and Spanger-Siegfried, E., eds. 2004. *Adaptation Policy Frameworks for Climate Change: Developing Strategies, Policies and Measures*. Cambridge, UK: Cambridge University Press.

MELAD (Ministry of Environment, Land and Agricultural Development) 2007. *National Adaptation Programma of Action: Kiribati*. Tarawa, Kiribati: Government of Kiribati.

MNREM (Ministry of Natural Resources, Environment and Meteorology) 2005. *National Adaptation Programma of Action: Samoa*. Apia, Samoa: Government of Samoa.

Morgan, M. and Dowlatabadi, H. 1996. Learning from integrated assessment of climate change. *Climatic Change*, **34**, 337–68.

Najam, A., Huq, S. and Sokona, Y. 2003. Climate negotiations beyond Kyoto: developing countries concerns and interests. *Climate Policy*, **3**(3), 221–31.

Nurse, L. and Sem, G. 2001. Small island states. In J. McCarthy, O. Canziani, N. Leary, D. Dokken and K. White, eds., *Climate Change 2001: Impacts, Adaptation and Vulnerability*. Cambridge, UK: Cambridge University Press, pp. 842–75.

O'Brien, K., Leichenko, R., Kelkar, U. *et al.* 2004. Mapping vulnerability to multiple stressors: climate change and globalisation in India. *Global Environmental Change*, **14**(4), 303–13.

Paavola, J., Adger, N. and Huq, S. 2006. Multifaceted justice I: adaptation to climate change. In N. Adger, J. Paavola, S. Huq and M. Mace, eds., *Fairness in Adaptation to Climate Change*. Cambridge, MA: MIT Press, pp. 263–78.

Paterson, M. 1996. *Global Warming and Global Politics*. London: Routledge.

Proctor, J. 1998. The meaning of culture in global environmental change: retheorizing culture in human dimensions research. *Global Environmental Change*, **8**, 227–48.

Rahmstorf, S. 2007. A semi-empirical approach to projecting future sea-level rise. *Science*, **315**(5810), 368–70.

Rayner, S. and. Malone, E. 1998. The challenge of climate change to the social sciences. In S. Rayner and E. Malone, eds., *Human Choice and Climate Change, Volume 4: What Have We Learned?* Ohio: Batelle Press, pp. 33–69.

Risbey, J., Kandlikar, M. and Patwardhan, A. 1996. Assessing integrated assessments. *Climatic Change*, **34**, 369–95.

Salinger, J. 2001. Climate variation in New Zealand and the Southwest Pacific. In A. Sturman and R. Spronken-Smith, eds., *The Physical Environment: A New Zealand Perspective*. Melbourne, Australia: Oxford University Press, pp. 130–49.

Schwartz, P. and Randall, D. 2003. *An Abrupt Climate Change Scenario and its Implications for United States National Security*. San Francisco, CA: Global Business Network. Available online: http://www.gbn.com/ArticleDisplayServlet.srv? aid=26231, accessed 8 April 2007.

Sen, A. 1999. *Development as Freedom*. New York: Anchor Books.

Shackley, S. and Gough, C. 2002. *The Use of Integrated Assessment: An Institutional Analysis Perspective*. Tyndall Centre for Climate Change Working Paper 14, Manchester.

Shackley, S., Young, P,, Parkinson, S. and Wynne, B. 1998. Uncertainty, complexity and concepts of good science in climate change modelling: are GCMs the best tools? *Climatic Change*, **38**, 159–205.

Shaw, M. 1993. There is no such thing as society: beyond individualism and statism in international security studies. *Review of International Studies*, **19**(2), 159–75.

Shibuya, E. 1996. Roaring mice against the tide: the South Pacific islands and agenda-building on global warming. *Pacific Affairs*, **69**(4), 541–55.

Smit, B., Burton, I., Klein, R. and Street, R. 1999. The science of adaptation: a framework for assessment. *Mitigation and Adaptation Strategies for Global Change*, **4**(3/4), 199–213.

Smit, B., Burton, I., Klein, R. and Wandel, J. 2000. An anatomy of adaptation to climate change and variability. *Climatic Change*, **45**(1), 223–51.

Smith, J., Schellnhuber, H. and Mirza, M. 2001. Vulnerability to climate change and reasons for concern: a synthesis. In J. McCarthy, O. Canziani, N. Leary, D. Dokken and K. White, eds., *Climate Change 2001: Impacts, Adaptation and Vulnerability*. Cambridge, UK: Cambridge University Press, pp. 914–67.

Sutherland, K., Smit, B., Wulf, V. and Nakalevu, T. 2005. Vulnerability in Samoa. *Tiempo*, **54**, 11–15.

Taylor, P. and Buttel, F. 1992. How do we know we have global environmental problems? Science and the globalization of environmental discourse, *Geoforum*, **23**, 405–16.

Thrift, N. 2004. Double geography. *Area*, **36**, 438–40.

Tschakert, P. 2007. Views from the vulnerable: understanding climatic and other stressors in the Sahel. *Global Environmental Change*, **17**, 381–96.

Tyler, N., Turi, J., Sundset, M. *et al.* 2007. Saami reindeer pastoralism under climate change: applying a generalized framework for vulnerability studies to a sub-arctic social–ecological system. *Global Environmental Change*, **17**(2), 191–206.

UNDP (United Nations Development Program) 1994. *Human Development Report 1994*. New York, NY: Oxford University Press.

UNFCCC (United Nations Framework Convention on Climate Change Secretariat) 2002. *Annotated Guidelines for the Preparation of National Adaptation Programmes of Action*. Bonn: LDC Expert Group, UNFCCC Secretariat. Available online: unfccc.int/ files/cooperation_and_support/ldc/application/pdf/annguide.pdf.

van Ireland, E., Klaassen, M., Nierop, T. and van der Wusten, H. 1996. *Climate Change: Socio-Economic Impacts and Violent Conflict*. Dutch National Research Programme on Global Air Pollution and Climate Change, Report No. 410 200 006, Wageningen.

Waever, O. 1995. Securitisation and desecuritisation. In R. Lipschutz, ed., *On Security*. New York: Columbia University Press, pp. 46–86.

Warrick, R., Kenny, G., Sims G., Ye, W. and Sem, G. 1999. The Vandaclim simulation model: a training tool for climate change vulnerability and adaptation assessment. *Environment, Development and Sustainability*, **1**(2), 157–70.

Woodward, A., Hales, S. and Weinstein, P. 1998. Climate change and human health in the Asia Pacific region: who will be most vulnerable? *Climate Research*, **11**(1), 31–8.

Yamin, F. 2004. Overview. *IDS Bulletin*, **35**(3), 1–10.

Part II

Equity

4

A "shared vision"? Why inequality should worry us

J. TIMMONS ROBERTS AND BRADLEY C. PARKS

Introduction: a "shared vision"?

In late 2007, the world sighed in relief after two grueling weeks of international climate negotiations that resulted in an upbeat-sounding 'Bali Roadmap.' The Roadmap identified a series of steps that might be taken to break the North–South impasse and solve the global climate crisis. In particular, a process under an Ad Hoc Working Group for Long-Term Cooperative Action under the Convention (AWG-LCA) was tasked with breaking the deadlock over who should act in cleaning up the atmosphere, and how. The answer, according to the Roadmap, was that developed and developing countries would move forward with "a shared vision for long-term cooperative action, including a long-term global goal for emissions reductions, to achieve the ultimate objective of the Convention [avoiding dangerous climate change]."

However, as negotiations moved on to Bonn, Accra, and Poznan in 2008, nearly every word of the Bali Action Plan was contested. In the run-up to the 14th Conference of the Parties of the UN Framework Convention on Climate Change (COP14) in Poznan, Poland, in December, 2008, 76 Parties submitted "Ideas and Proposals" to the Working Group. China asserted that developed countries would need to "tak[e] the lead in reducing their emissions of greenhouse gases, while ensuring development rights and spaces for developing countries." Only with such a mid-term target being clearly determined, they argued, is it meaningful to talk about any long-term goals for emission reductions (UNFCCC, 2008b). The G-77 called for the creation of a "Financial Mechanism for Meeting Financial Commitments under the Convention" to force rich nations to honor their many promises of assistance. China also argued that total assistance should amount to 0.5–1% of the annual GNP of Annex I Parties and emphasized any assistance related to a post-2012 climate treaty should be additional to the existing official development assistance.

Climate Change, Ethics and Human Security, eds. Karen O'Brien, Asunción Lera St.Clair and Berit Kristoffersen. Published by Cambridge University Press. © Cambridge University Press 2010.

Country after country stated the need for atmospheric clean-up actions to be guided by the principle of "common but differentiated responsibilities and respective capabilities" (UNFCCC, 2008a). Brazil argued that "an important equity factor ... will be that ... countries should contribute to the solution according to their contribution to the problem." Brazilian representatives also noted that "developed countries ... should demonstrate the leadership required ... [and] must achieve absolute reductions. Developing countries, despite their limited historical responsibility for climate change, face the highest costs regarding [climate change] impacts" (UNFCCC, 2008a).[1]

Therefore, as we turn the page to a new era of North–South climate negotiations, the issue of equity, first raised in Stockholm in 1972 and brought up at every major international environmental summit since then, continues to cast a long shadow over efforts to forge an effective global climate agreement. Developing countries are generally not willing to protect the global environment if they feel that other countries with higher levels of responsibility and greater capacity to act are not making good faith efforts to address the issue. Yet a global climate agreement without Southern participation is of little value: Kyoto is only binding for a group of countries that account for 19% of global emissions. These "Annex 1" countries are required to reduce their emissions by roughly 5%, which will likely have little impact on climate stability. At the same time, developing countries will likely be responsible for roughly 60% of global emissions by 2030. In this chapter, we argue that the stalemate in North–South climate negotiations is unlikely to be resolved in the absence of aggressive efforts to address issues of inequality and justice.

In 1994, The United Nations Development Program identified a series of threats to human security, which included environmental disasters and threats to human wellbeing associated with climate change. To address the human security issues raised by climate change, we argue that the international community must pay close attention to inequality not only in who is emitting the most greenhouse gases, and which countries are most vulnerable to climate change impacts, but in the global distribution of wealth and power.

We first review the broad dimensions of inequality related to climate change – in vulnerability, responsibility, and action. We then turn to how these inequalities have influenced international environmental negotiations over the last few decades. Promises of funding have helped bring about sporadic progress on the issue, but we consider increased funding as part of a broader and more ambitious effort to

---

[1] The Third World Network questioned the whole exercise, pointing out that "the issue of long term goals has major development and equity implications ... developing countries could be committing themselves to a cut of certain percentage in their emissions without directly being aware of this" (TWN, 2008, http://www.twnside.org.sg/title2/climate/pdf/TWN%20submission_global%20goal.pdf).

integrate climate change and development, and rebuild the social trust necessary to develop a truly shared vision of global climate policy.

The corrosive impact of inequality on North–South global climate negotiations

The absence of an effective global climate treaty 20 years after the problem was identified, in spite of increasingly dire scientific evidence, raises broader questions about the factors that shape international environmental cooperation. Scholars and policy analysts have identified a broad range of factors that seem to influence outcomes in international environmental politics: material self-interest; bargaining power; international rules, norms, and decision-making procedures; non-state actors, such as epistemic communities, NGOs, and corporations; crises; political leadership; and domestic political institutions (Haas, 1990; Sprinz and Vaahtoranta, 1994; Young, 1994; Wapner, 1995; Victor, 2001; Roberts and Parks, 2007). Yet interestingly, one of the variables often singled out by Southern policy makers as a major impediment to cooperation – global inequality – has not received much scholarly attention.[2]

Inequality, we argue, can dampen utility-enhancing cooperative efforts by reinforcing "structuralist" worldviews and causal beliefs, polarizing policy preferences, making it difficult to coalesce around a socially shared understanding of what is "fair," eroding conditions of trust, generating divergent and unstable expectations about future behavior, and creating incentives for zero-sum and negative-sum behavior. In this chapter, we quickly review three main sources of inequality: responsibility for the problem; vulnerability to climate-related shocks and stresses; and uneven participation in global efforts to solve the problem. We then examine some of the different channels through which inequality may negatively influence the prospects for North–South cooperation. We conclude by exploring several policy options and providing historical examples that illustrate how countries with highly disparate worldviews, causal beliefs, principled beliefs, and policy positions have resolved their differences and cooperated on issues of mutual interest.

Responsibility

With only 4% of the world's population, the United States is responsible for over 20% of all global emissions. That can be compared to 136 developing countries that together are only responsible for 24% of global emissions (Roberts and Parks, 2007). Poor countries therefore remain far behind wealthy

---

[2] There are, of course, a few noteworthy exceptions (see Müller, 1999; Najam, 2004; Chasek et al., 2006).

countries in terms of emissions per person. Overall, the richest 20% of the world's population is responsible for over 60% of its current emissions of greenhouse gasses. That figure surpasses 80% if past contributions to the problem are considered, and they probably should be, since carbon dioxide, the main contributor to the greenhouse effect, often remains in the atmosphere for over 100 years.

However, there are many ways to understand emissions inequality and responsibility for climate change, and each approach represents a different social understanding of fairness. Four of these approaches are: (1) grandfathering (i.e., that countries should reduce from a baseline year, such as 1990, which was the basis for the Kyoto Protocol) falls in line with the entitlement principle, whereby individuals are entitled to what they have or have produced; (2) the carbon intensity approach, which is usually associated with a measure of CO_2 emissions per unit of GDP, represents the utilitarian principle that inefficient solutions are also unjust since everyone is worse off in the absence of joint gains; (3) the historical responsibility approach focuses on how much countries have contributed to the stock of greenhouse gases in the atmosphere, which is the basis for "the polluter pays" principle; (4) the equal emissions rights per capita approach is consistent with the egalitarian principle that every human should have equal rights to global public goods, such as atmospheric stability. These different perceptions of fairness are to a large extent shaped by the highly disparate positions that countries occupy in the global hierarchy of economic and political power. In this way, we argue that inequality has a dampening effect on cooperation by polarizing policy preferences and making it difficult for countries to arrive at a socially shared understanding of what is "fair."

Vulnerability

The scientific community agrees that carbon emissions will create a warmer and often wetter atmosphere, and, in turn, increase flooding, hurricanes, forest fires, winter storms, and drought in arid and semi-arid regions. Climatologists have observed a sharp upswing in the frequency, magnitude, and intensity of hydro-meteorological disasters over the past two decades – the five warmest years on historical record were 1998, 2002, 2003, 2005, and 2007 – and hydrometeorological disasters have more than doubled since 1996 (Goddard Institute for Space Studies, 2008).

Although climate change is often characterized as "everybody's problem," hydrometeorological impacts are socially distributed across human populations (Kaul *et al.*, 1999). Some countries and communities will suffer more immediately and profoundly, and they are generally not those most responsible for creating the problem (Roberts and Parks, 2007). According to the latest predictions of the Inter-governmental Panel on Climate Change (IPCC), rapidly expanding populations in Africa, Asia, and Latin America are suffering disproportionately from more frequent and dangerous droughts, floods, and storms (IPCC, 2007). The World Bank reports that "[b]etween 1990 and 1998, 94% of the world's disasters and 97% of all natural-disaster-related deaths occurred in developing countries" (Mathur *et al.*, 2004: 6). In relative terms, ten times more, and in some cases hundreds of times more, people are dying in the developing world than in the United States, the United Kingdom, and other Western countries. For example, in the United States, less than one-seventh of one percent of the population was made homeless by hydrometeorological disasters between 1980 and 2002. By contrast, in Bangladesh 45 percent of the population was at some point made homeless by a climate-related disaster during the same period.

Unequal vulnerability to climate change may influence the prospects for North–South cooperation: poor countries suffering from rising sea levels, devastating droughts and storms, lower agricultural yields, and increased disease burdens are unlikely to be enthusiastic about cleaning up an environmental problem that the industrialized world created in the first place.[3] Indeed, stark inequalities in vulnerability have already poisoned the negotiating atmosphere. "If climate change makes our country uninhabitable," said Bangladeshi Atiq Rahman during the 1995 Berlin negotiations, "we will march with our wet feet into your living rooms" (Athanasiou and Baer, 2002: 23). At every subsequent COP, developing countries have underscored their small contribution to the problem of climate change and their extreme vulnerability to its impacts (Müller, 2001).[4] While some climate policy analysts dismiss this line of argumentation as mere posturing, a 2008 European Union report warns that "[c]limate change impacts will fuel the politics of resentment between those most responsible for climate change and those most affected by it" (European Union, 2008: 5).

Action

There are also stark inequalities in who is currently doing something to reduce greenhouse gas emissions, and which countries will likely bear the greatest burden of atmospheric clean-up in the future. Although Northern governments are trying to convince the Southern governments that they need to rein in their greenhouse gas

[3] Conversely, one might argue that self-interest would make more vulnerable countries more likely to join global efforts to reduce greenhouse gas emissions; see Sprinz and Vaahtoranta (1994).

[4] Fifteen years ago, Young (1994: 50) noted that "[s]ome northerners may doubt the credibility of [threats from southern nations to damage the global climate] and advocate a bargaining strategy that offers few concessions to the developing countries. But such a strategy is exceedingly risky. Many of those located in developing countries are increasingly angry and desperate . . . Faced with this prospect, northerners will ignore the demands of the South regarding climate change at their peril."

emissions, most of them are not doing so in their own countries. Under the Kyoto Protocol, "Annex I" (developed) countries committed to a 5.2% (average) reduction in greenhouse gas emissions (below 1990 levels) by 2012. However, with the exception of several European countries, greenhouse gas emissions have risen significantly throughout the industrialized world since 1990. Even President Barack Obama's pledge to reverse US policy under President George W. Bush has only resulted in a promise to reduce US emissions by 17% from 2005 levels through 2020, back to just 3% below 1990 levels. Meanwhile, following the IPCC's 2007 recommendations, China has called on "all developed country Parties to the Convention [to] commit to a reduction in GHG emissions by at least 25–40 percent below 1990 levels in 2020 and by approximately 80–95% in 2050" (AWG-LCA contribution, September 28, 2008).

At the COP14 meeting in Poznan, the EU seemed to be split between its Eastern and Western member states over whether it can meet those targets. Moreover, many industrialized countries have indicated that rather than making cuts at home, they would prefer to achieve their emission reduction commitments by funding activities in developing countries. From a cost-efficiency perspective, this makes good sense: the greatest opportunities for low-cost emissions reductions exist in the developing world (Stavins and Olmstead, 2006). However, there are many moral and practical problems with the rich merely paying the poor to do the cleaning up for them. Simply stated, the "demandeurs" of global climate protection face a credibility problem: they need to demonstrate that they are willing to make difficult choices at home before they can enlist the support of developing countries.

The last 35 years of global environmental negotiations have demonstrated that developing countries have deeply held distributional concerns, which can be a significant impediment to international cooperation. According to one "Group of 77" expert, the South's "principal fear . . . [is] that the North is using environmental issues as an excuse to pull up the development ladder behind it – [a suspicion which] has remained unallayed through two decades of environmental diplomacy" (Najam, 1995: 249). Joanna Depledge (2002), a former UNFCCC Secretariat staff member, has similarly reported that many non-Annex I (developing) countries fear efforts to curb carbon emissions in the developing world will effectively place a "cap" on their economic growth.[5]

It is also important to note that even among developed countries that appear to have reduced or stabilized their greenhouse gas emissions since 1990, there are serious questions about whether such national statistics on greenhouse gas emissions truly indicate a shift from high-carbon to low-carbon economies and lifestyles. New

---

[5] During the COP14 Poznan negotiations, one could also perceive this concern in the language of the so-called "African Group," which stressed that "a shared vision also involves sustainable development" (ENB, December 2, 2008).

research suggests that many "service-exporting" OECD countries, which specialize in areas like banking, tourism, advertising, sales, product design, procurement, and distribution, are in many cases "net-importers" of carbon-intensive goods coming primarily from developing countries. As such, they do not necessarily emit less; they may simply displace their emissions (Heil and Selden, 2001; Machado *et al.*, 2001; Muradian *et al.*, 2002).[6] This changing pattern of production and consumption has not gone unnoticed. In 2008, Chinese Minister of Foreign Affairs, Yang Jiechi, pointed out that many of China's carbon emissions are the by-product of Northern demand for manufactured goods, stating "I hope when people use high-quality yet inexpensive Chinese products, they will also remember that China is under increasing pressure of transfer emission[s]" (Economic Times, 2008).

Inequality and mistrust in international environmental regimes

International climate negotiations are deeply embedded in the broader context of North–South relations. In 1972, at the first international conference on the environment in Stockholm, Sweden, it quickly became evident that no consensus would emerge between developed and developing countries on the issue of global environmental protection. "Late developers" feared restrictions on their economic growth, emphasized the North's profligate use of planetary resources, and pushed for a redistributive programme that would benefit them economically and hasten the transition towards industrialization. Developed countries wanted Northern consumption off the negotiating table, Southern population growth on the agenda, and non-binding language on issues of financial assistance and technology transfer (Haas *et al.*, 1993). Neither negotiating bloc was willing to budge, and deeply held feelings of marginalization and injustice among poor nations made for an adversarial negotiating atmosphere.

The South's confrontational approach intensified in the late 1970s under the banner of the "New International Economic Order" (NIEO). During this period, developing countries put forth a "series of proposals ... which included significant wealth redistribution, greater LDC participation in the world economy, and greater Third World control over global institutions and resources" (Sebenius, 1991: 128). At the same time, late developers became strident in their criticism of Northern environmentalism – an environmentalism that many perceived as "pull[ing] up the development ladder" (Najam, 1995).

---

[6] This pattern has been celebrated as a clear indication that rich countries are becoming more "post-materialist" or "post-industrial." However, these service economies still require extraordinary levels of energy and materials. Substantial research has demonstrated that the production of the material goods (and their effluents) has shifted over time to poorer countries. As such, the material-intensive imports required by rich countries have carbon emissions "embodied" within them (Machado *et al.*, 2001).

In further rounds of negotiations, on issues such as biodiversity, desertification, and climate change, there were calls for increased financial compensation and more equitable representation (DeSombre and Kaufman, 1996; Sell, 1996). Debate over the voting structure of the Global Environmental Facility, which distributes hundreds of millions of dollars of environmental aid each year, became especially conflict ridden. Poor and middle-income countries protested "donor dominance" and the lack of transparency in decision making, while rich, industrialized countries insisted that only the "incremental costs" of environmental projects with global benefits be financed (Keohane and Levy, 1996). North–South environmental relations also suffered several important setbacks after the 1992 Rio Earth Summit (Raustiala, 1997). Developed countries agreed to underwrite the participation of less developed countries in any global environmental accord to come out of Rio. Specifically, they agreed to a financial package of 100 billion US dollars (USD) a year in new and additional concessionary funds for "sustainable development" and USD 15 billion for global environmental issues (Robinson, 1992). However, wealthy OECD countries failed to honor their policy commitments (Najam, 2002). At the end of the 1990s, less than ten billion USD a year was being allocated for environmental issues, about 20 percent of the Rio promise (Hicks *et al.*, 2008). The reasons for the unmet promises varied: recessions at home, new electoral coalitions in power, executive commitments that legislatures refused to ratify or sustain, or simple backsliding.

Then in 1997, at the UN General Assembly Special Session for Review and Appraisal of Agenda 21 (UNGASS), developing countries sought to strengthen the sustainable development agenda by linking the issues of climate change, forests and biodiversity to issues of trade, investment, finance, and intellectual property rights. This was flatly rejected by rich nations (Sandbrook, 1997). Three years later, at the COP6 climate negotiations, developing country delegations expressed outrage after the (Western) chairs allegedly deleted text that had been agreed upon earlier (Dessai, 2001). The G77 and China also charged that many of the important decisions affecting developing countries were being made in non-transparent "Green Room" meetings, attended only by powerful countries. This set the stage for the 2002 World Summit on Sustainable Development (WSSD), where one reporter noted that "effective governance is not possible under the prevailing conditions of deep distrust" (Najam, 2003: 370). As seen in Copenhagen in 2009 and as we argue below, this lack of trust in North–South relations has proven to be a major obstacle to the creation of a post-2012 global climate pact.

International climate negotiations are also inextricably linked to North–South *economic* relations. Stephen Krasner once said that there are "makers, breakers, and takers" in international relations, and there is little question that developing countries are generally "takers" in international economic regimes (Krasner, 1978). "[T]he 'price' of multilateral rules," explains Shadlen, "is that [Least Developed

Countries – LDCs] must accept rules written by – and usually for – the more developed countries" (Shadlen, 2004: 6). Gruber (2000) argues that powerful states – particularly those with large markets – possess "go-it-alone power" in that they can unilaterally eliminate the previous status quo and proceed gainfully with or without the participation of weaker parties.

Robert Wade refers to a so-called "shrinking of development space," and argues that "the rules being written into multilateral and bilateral agreements actively prevent developing countries from pursuing the kinds of industrial and technology policies adopted by the newly developed countries of East Asia and by the older developed countries when they were developing" (Wade, 2003: 622). Similarly, Birdsall *et al.* (2005) explain how the callous – and at times opportunistic – actions of Western governments have made upward mobility in the international division of labour difficult. Other scholars of international political economy have highlighted the fact that the governance structures of international financial institutions, like the International Monetary Fund and World Bank, prevent the institutions' main clients (developing countries) from having any significant voting power (Woods, 1999; Wade, 2003).

These inequalities of opportunity have an indirect, but important, impact on how developing countries approach global environmental negotiations. Porter and Brown (1991: 124; see also Chasek *et al.*, 2006) find that "developing states' perceptions of the global economic structure as inequitable has long been a factor in their policy responses to global environmental issues." Similarly, Gupta (2000) has reported that "[Southern] negotiators tend to see issues holistically and link the issue to all other international issues. Thus linkages are made to international debt, trade and other environmental issues such as desertification." As we have argued elsewhere (e.g. Roberts and Parks, 2007), when powerful states disregard weaker states' positions in the international division of labour in areas where they possess structural power (as in international economic regimes), they run a high risk of weaker states "reciprocating" in policy areas where they possess more bargaining leverage (as in international environmental regimes).

Looking ahead towards solutions

We now return to the current North–South impasse over climate change with which we began. The Kyoto Protocol requires emission reduction commitments from a group of wealthy countries that account for less than one-fifth of global carbon emissions. These countries are required to reduce their emissions by a small percentage, which will likely have almost no impact on atmospheric stability. At the same time, developing country emissions are expected to skyrocket – to 60% of total global emissions –over the next 20 years. Therefore, while the Kyoto Protocol may have been useful from a political perspective, it is clear that negotiators must now focus on the central task of enlisting the active participation of developing countries in a "post-2012" global climate regime. As Wheeler and Ummel (2007: 10) put it in no uncertain terms, climate stabilization will demand that "the South . . . accept the necessity of serious, costly mitigation, and immediately embark on a low-carbon development path."[7]

However, given the South's record of vigorously resisting any binding commitments related to future emissions reductions, there is a great deal of uncertainty about how to most effectively engage developing countries going forward. Some climate policy analysts have argued that negotiators should soldier on and continue with the same kind of horse-trading tactics that have characterized the first 20 years of climate negotiations. According to this view, widening the scope of international negotiations to include issues like inequality, trust, and fairness will be the millstone around the neck of any effort to stabilize the climate. It is instead thought that Western governments should invest in clean technologies, de-link economic growth from carbon emissions, strengthen compliance mechanisms, and proceed with or without the participation of a majority of developing countries. One might call this the "pragmatic justice" approach – the approach that says a perfectly fair agreement existing only in the minds of negotiators is in fact unfair to all parties. But the "pragmatic justice" approach overlooks the daunting scientific reality that the world is already on track to seriously destabilize the atmosphere by 2020 or 2030, and it ignores the South's extreme risk aversion to binding emission limits.

We argue that breaking the North–South impasse on global climate policy will likely require unconventional – and perhaps even heterodox – policy interventions. To date, countries have proposed different yardsticks for measuring atmospheric clean-up responsibilities based on particularistic notions of justice (Roberts and Parks, 2007). But high levels of inequality make it very unlikely that a North–South consensus will spontaneously emerge on the basis of a single fairness principle. Therefore, a moral compromise, or "negotiated justice" settlement, will almost certainly be necessary. To break through the cycle of mistrust that plagues North–South relations, we also argue that the North needs to offer the South a new global bargain on environment and development and signal its commitment to this new 'shared vision' through a series of confidence building measures.

Moving towards "hybrid justice"

Earlier, we mentioned four very different approaches to measuring national responsibility for greenhouse gas emissions: the grandfathering approach, which relies on entitlement principles of justice; the carbon intensity approach, which rests on

---

[7] The first commitment period under the Kyoto Protocol is set to expire in 2012.

utilitarian principles of justice; the historical responsibility approach, which oper-
ationalizes the "polluter pays" principle; and equal rights emissions per capita
approach. Since these particularistic notions of justice are closely associated with
where countries sit in the global hierarchy of economic and political power, it is
unlikely that a North–South fairness consensus will spontaneously emerge on the
basis of one of these principles. Instead, we argue that a moral compromise, or
"negotiated justice" settlement, is necessary.

In recent years, a number of proposals representing moral compromise have
emerged. Bartsch and Müller (2000) propose a "preference score" method, which
combines the grandfathering and per capita approach through a voting system. Their
proposal allows each country – weighted by their population – to choose the
methodology that they prefer. Each global citizen's "vote" is then used to calculate
national carbon emission allowances. According to their preliminary model, under
this proposal, roughly three-quarters of the global emissions budget would be based
on the per capita approach and one-quarter on grandfathering. Others have focused
on more politically feasible per capita proposals that provide for "national circumstances," or allowance factors, like geography, climate, energy supply, and domestic
economic structure, as well as "soft landing scenarios" (e.g. Gupta and Bhandari,
1999; Ybema *et al.*, 2000; Agarwal *et al.*, 2001; Baumert and Kete, 2002; Ringius
et al., 2002; Torvanger and Ringius, 2002; Torvanger and Godal, 2004).

The Pew Center for Global Climate Change has developed a hybrid proposal that
assigns responsibility based on past and present emissions, carbon intensity and
countries' ability to pay (that is its per capita GDP) (Claussen and McNeilly, 1998).
It separates the world into three groups: those that "must act now," those that "could
act now," and those that "should act now, but differently." The "Triptych" proposal,
designed by scholars at the University of Utrecht (and already used differentiate
commitments among EU countries), "accounts for differences in national circumstances such as population size and growth, standard of living, economic structure
and fuel mix in power generation" (Groenenberg *et al.*, 2001). Its novel contribution
is that it divides each country's economy into three sectors: energy-intensive
industry, power generation, and the so-called domestic sector (transport, light
industry, agriculture, and commercial sector). It applies the carbon intensity
approach to the energy-intensive sector, "decarbonization targets" to the power
generation sector and a per capita approach to the "domestic" sectors. Similarly,
the multisector convergence approach, developed by two research institutes in
northern Europe (ECN and CICERO), treats sectors differentially and integrates
per capita, carbon intensity, and ability to pay (GDP per capita) approaches (Sijm
et al., 2000).

A small group called EcoEquity.org has also created a "Greenhouse Development
Rights" framework as a reference to evaluate proposals for the post-2012 period

(Baer *et al.*, 2008). They argue that individuals below a "global middle-class" income of USD 7 500 per capita should be assured that they will not be asked to make binding limits until they approach that level, while countries above that level should be responsible for rapid reductions of emissions and payments to assist those below the line in improving their social and economic status while adjusting to a less carbon-intensive path of development. Funds raised in wealthy countries in reducing emissions are used to help poor countries adapt and develop in a more climate-friendly way. We believe these hybrid proposals are among the most promising solutions to break the North–South stalemate.

Building trust and developing a "shared vision"

Drawing upon Andrew Kydd's research on US–Soviet relations in the run-up to the end of the Cold War, we would argue that a series of costly signals – "signals designed to persuade the other side that one is trustworthy by virtue of the fact that they are so costly that one would hesitate to send them if one were untrustworthy" (Kydd, 2000: 326) – can foster mutual trust between countries that do not have a long history of cooperation. These measures should offer a new vision of global environmental cooperation, provide opportunities for developing countries to transition towards less carbon-intensive development pathways and clearly signal a desire to reverse long-standing patterns of global inequality. Additionally, we emphasize the central importance of exercising self-restraint when the short-term payoff on opportunistic behaviour is high. When powerful states consistently treat weaker states like second-class citizens, they run the risk of weaker states "reciprocating" in policy domains where they possess greater bargaining leverage.

The conditions of mistrust which currently plague North–South environmental relations can be understood as the product of a "failed reassurance strategy." In the early 1990s, the North assured poorer nations that they would "take the lead" in stabilizing the climate. However, subsequent efforts have been sluggish, litigious, uneven, and generally unimpressive. The lack of progress by the United States and other industrialized nations in meeting their own emission reduction targets has provided developing nations a ready excuse for not making cuts. As Brazil's leading newspaper put it in the era of Kyoto's signing, "[n]umbers like these [the US's emissions] reinforce the disposition of the Brazilian government to reject the idea of taking on additional costs to do its part in reducing the greenhouse effect" (Rossi, 1997).

However, there are some examples of modestly successful trust-building efforts in global environmental politics. The Multilateral Ozone Fund enshrined the "compensatory justice" principle and gave developing countries a greater stake in the decision-making process governing the allocation of environmental aid (Woods,

1999; Agarwal et al., 2001; Hicks et al., 2008). The Montreal Protocol also gave developing countries a 10-year window to pursue "cheap" economic development before making serious chlorofluorocarbon reductions.

Sometimes trust-building is also about exercising strategic restraint. As we have argued elsewhere, one important way to send a "costly signal" would be to aggressively support the interests and priorities of developing countries in the international political economy arena (Roberts and Parks, 2007). In fact, this could ultimately prove to be *more* important than international treaties, carbon accounting schemes, or environmental aid. According to seasoned analyst Herman Ott and others, "it became clear [at COP8 in New Delhi] that developing countries would not give up their 'right' for increasing emissions without serious concessions in other fields of the development agenda which satisfy the demand for global equity and poverty reduction" (Ott, 2004: 261).

Scholars of environmental politics unfamiliar with the international political economy literature may view such demands as distracting and unconstructive, but the ongoing development crisis is at the very heart of the climate policy gridlock. Developing countries want more "policy space" – room to define and pursue their own development agenda – but contemporary international economic regimes present huge hurdles to export diversification, institutional experimentation, and upward mobility in the world economy (Wade, 2003).

As such, if industrialized nations are interested in securing a North–South global climate agreement, we argue that they should consider explicitly signaling their concern for the "structural obstacles" facing developing countries. For example, the current practice of tariff escalation reinforces the structuralist perception that rich countries do not want poor countries to get rich the same way they did. The WTO's intellectual property agreement, TRIPs (The Agreement on Trade-Related Aspects of Intellectual Property Rights) has a similar effect since rich nations historically had complete policy autonomy in this area, granting patents at their own discretion in order to encourage industrial transformation (Shadlen, 2004; Birdsall et al., 2005: 8). Other possibilities include reining in the "deep integration" and anti-industrial policy crusade, not punishing poor countries for export diversification efforts, recognizing that the "political losers" created by the diversification process must be somehow compensated and promoting predictability (and reducing opportunities for opportunism) in international economic regimes (Wade, 2003; Shadlen, 2004; Birdsall et al., 2005).

We do recognize, however, that under circumstances of extreme mistrust risk-averse states may require more than costly signals and strategic reassurance. Indeed, it may be necessary for would-be co-operators to work towards establishing a "shared world view" or "shared vision." Kydd (2000) argues that the creation of a shared world view is typically the result of one state (or group of states) trying to get another state (or group of states) to buy into a so-called "new thinking." For example, during the Cold War, the US and Soviet administrations worked together to establish a "new thinking" about global security. When the Soviet Union withdrew from Afghanistan, an editorial appeared in *The New York Times*, noting that its actions "begin to render credible Moscow's "new thinking" about the Soviet role in the world" (quoted in Kydd, 2000: 346). Subsequently, when "asked if he still held to the idea that the Soviet Union was an evil empire . . . [Reagan] responded, 'No, I was talking about another time, another era.'"

Athanasiou and Baer (2002: 83) suggest that, in the context of international climate negotiations, the greatest challenge is to ensure that the South "not [view] climate justice as the justice of following the North down the fossil-fuel path." Wheeler and Ummel (2007: 9) similarly note that policy makers need to be disabused of "the notion that the South can utilize carbon-intensive growth to dramatically increase incomes – a kind of last-minute, fossil-fuelled development push – before the onset of catastrophic climate change." But to move away from "old North–South thinking," or what Graham (1996: 216) calls "residual 1970s thinking," an attractive alternative must be offered. We argue that the North will need to aggressively assist developing countries in making the tough transition to lower carbon pathways of development. This is not merely a financing issue. Countries on high-emission pathways will require serious attention to their political and class situations, as diversification is an intensely political process and conflict will inevitably arise. Developing countries will therefore need "policy space" to pursue strategies tailored to local culture, knowledge, institutions, and politics, while being provided significant technical assistance, technology transfer, and aid.

To conclude, climate change is fundamentally an issue of inequality and its resolution will likely demand an unconventional policy approach. Climate negotiations, we must remember, take place in the context of an ongoing development crisis, and what is perceived by the global South as a pattern of Northern callousness and opportunism in matters of international political economy. Copenhagen confirms that they take place at a time when levels of generalized trust are declining. And they take place at a time when poor nations' concerns for fair processes and outcomes have frequently been marginalized. Negotiators must therefore explicitly and aggressively signal concern and seek to address the structural obstacles facing developing countries. We need a global and just transition built on diffuse reciprocity, a climate of trust, negotiated justice, and a shared vision of truly long-term cooperative action.

References

Agarwal, A., Narain, S., Shama, S. and Imchen, A. 2001. *Poles apart: global environmental negotiation-2*. New Delhi, India: Centre for Science and Environment.

Athanasiou, T. and Baer, P. 2002. *Dead Heat: Global Justice and Global Warming*. New York, NY: Seven Stories Press.

Baer, P., Athanasiou, T., Kartha, S. and Benedict, E. K. 2008. *The Greenhouse Development Rights Framework: The Right to Development in a Climate Constrained World. Publication Series on Ecology. Vol. 1*. 2nd edn. Berlin: Heinrich BoÅNII Foundation, Christian Aid, Eco Equity and the Stockholm Environment Institute. Available online: gdrights.org/wp-content/uploads/2009/01/the gdrsframework.pdf.

Bartsch, U. and Müller, B. 2000. *Fossil Fuels in a Changing Climate: Impacts of the Kyoto Protocol and Developing Country Participation*. Oxford, UK: Oxford University Press.

Baumert, K. A. and Kete, N. 2002. An architecture for climate protection. In Kevin Baumert, eds., *Building on the Kyoto Protocol: Options for Protecting the Climate*. Washington, DC: World Resources Institute, pp. 1–30.

Birdsall, N., Rodrik, D. and Subramanian, A. 2005. *If Rich Governments Really Cared About Development*. Working Paper, Geneva: International Centre for Trade and Sustainable Development.

Chasek, P., Downie, D. and Brown, J., W. 2006. *Global Environmental Politics*, 4th edn. Boulder, CO: Westview Press.

Claussen, E. and McNeilly, L. 1998. *Equity and Global Climate Change: The Complex Elements of Fairness*. Arlington, VA: Pew Center on Climate Change.

Depledge, J. 2002. Continuing Kyoto: extending absolute emission caps to developing countries. In K. Baumert, ed., *Building on the Kyoto Protocol: Options for Protecting the Climate*. Washington, DC: World Resources Institute, pp. 31–60.

DeSombre, E. R. and Kaufman, J. 1996. The Montreal Protocol Multilateral Fund: Partial Success Story. In R. O. Keohane and M. A. Levy, eds., *Institutions for Environmental Aid: Pitfalls And Promise*. Cambridge, MA: MIT Press, pp. 89–126.

Dessai, S. 2001. Why did The Hague climate conference fail? *Environmental Politics*, **10**(3), 139–44.

Economic Times 2008. China tells developed world to go on climate change "diet". *Economic Times*, March 12, 2008.

European Union. 2008. *Climate Change and International Security*. Paper from the High Representative and the European Commission to the European Council, Brussels, March 14. Available online: www.consilium.europa.eu/ueDocs/cms_Data/docs/pressData/en/reports/99387.pdf.

Goddard Institute for Space Studies 2008. Global temperature anomalies in. 01°C. Available online: http://data.giss.nasa.gov/gistemp/tabledata/GLB.Ts.txt, accessed July 10, 2008.

Graham, E. M. 1996. Direct investment and the future agenda of the World Trade Organization. In J. J. Schott, eds., *The World Trading System: Challenges Ahead*. Washington, DC: Institute for International Economics, pp. 205–17.

Groenenberg, H., Phylipsen, D. and Blok, K. 2001. Differentiating commitments worldwide: global differentiation of GHG emissions reductions based on the Triptych approach. A preliminary assessment. *Energy Policy*, **29**(12), 1007–30.

Gruber, L. 2000. *Ruling the World: Power Politics and the Rise of Supranational Institutions*. Princeton, NJ: Princeton University Press.

Gupta, J. 2000. *On Behalf of my Delegation: A Survival Guide for Developing Country Climate Negotiators*. Washington, DC: Center for Sustainable Development in the Americas.

Gupta, S. and Bhandari, P. M. 1999. An effective allocation criterion for CO_2 emissions. *Energy Policy*, **27**(12), 727–36.

Haas, P. M. 1990. *Saving the Mediterranean: The Politics of International Environmental Cooperation*. New York, NY: Columbia University Press.

Haas, P., Keohane, R. and Keohane, M., eds. 1993. *Institutions for the Earth: Sources of Effective International Environmental Protection*. Cambridge, MA: MIT Press.

Heil, M. T. and Selden, T. M. 2001. International trade intensity and carbon emissions: a cross-country econometric analysis. *Journal of Environment and Development*, **10**(1), 35–49.

Hicks, R. L., Parks, B. C., Roberts, J. T. and Tierney, M. J. 2008. *Greening Aid? Understanding the Environmental Impact of Development Assistance*. Oxford, UK: Oxford University Press.

IPCC 2007. *Climate Change 2007: Fourth Assessment Report of the Intergovernmental Panel on Climate Change*. Cambridge, UK: Cambridge University Press.

Kaul, I., Grunberg, I. and Stern, M. 1999. Defining global public goods. In I. Kaul, I. Grunberg and M. Stern, eds., *Global Public Goods: International Cooperation in the 21st Century*. Oxford, UK: Oxford University Press, pp. 2–19.

Keohane, R. and Levy, M. A., eds. 1996. *Institutions for Environmental Aid: Pitfalls and Promise*. Cambridge, MA: MIT Press.

Krasner, S. 1978. United States commercial and monetary policy: unraveling the paradox of external strength and internal weakness. In P. J. Katzenstein, eds., *Between Power and Plenty: Foreign Economic Policies of Advanced Industrial States*. Madison, WI: University of Wisconsin Press, pp. 51–87.

Kydd, A. 2000. Trust, reassurance, and cooperation. *International Organization*, **54**(2), 325–57.

Machado, G., Schaeffer, R. and Worrell, E. 2001. Energy and carbon embodied in the international trade of Brazil: an input-output approach. *Ecological Economics*, **39**(3), 409–24.

Mathur, A., Burton, I. and van Aalst, M., eds. 2004. *An Adaptation Mosaic: A Sample of the Emerging World Bank Work in Climate Change Adaptation*. Washington, DC: The World Bank.

Müller, B. 1999. *Justice in Global Warming Negotiations: How to Obtain a Procedurally Fair Compromise*. Oxford, UK: Oxford Institute for Energy Studies.

Müller, B. 2001. *Fair Compromise in a Morally Complex World*. Paper presented at Pew Equity Conference, Washington, DC, April 17–18, 2001.

Muradian, R., O'Connor, M. and Martinez-Alier, J. 2002. Embodied pollution in trade: estimating the "environmental load displacement" of industrialized countries. *Ecological Economics*, **41**(1), 51–67.

Najam, A. 1995. International environmental negotiations: a strategy for the South. *International Environmental Affairs*, **7**(2), 249–87.

Najam, A. 2002. Financing sustainable development: crises of legitimacy. *Progress in Development Studies*, **2**(2), 153–60.

Najam, A. 2003. The case against a new international environmental organization. *Global Governance*, **9**(3), 367–84.

Najam, A. 2004. The view from the South: developing countries in global environmental politics. In R. Axelrod, D. Downie and N. Vig, eds., *The Global Environment: Institutions, Law, and Policy*, 2nd edn. Washington, DC: CQ Press, pp. 225–43.

Ott, H. 2004. Global climate. In G. Ulfstein and J. Werksman, eds., *Yearbook of International Environmental Law 12*. Oxford, UK: Oxford University Press, pp. 261–70.

Porter, G. and Brown, J. W. 1991. *Global Environmental Politics*. Boulder, CO: Westview Press.

Raustiala, K. 1997. Domestic institutions and regulatory cooperation: comparative responses to the Global Biodiversity Regime. *World Politics*, **49**(4), 482–509.

Ringius, L., Torvanger, A. and Underdal, A. 2002. Burden sharing and fairness principles in international climate policy. *International Environmental Agreements: Politics, Law and Economics*, **2**(1), 1–22.

Roberts, J. T. and Parks, B. C. 2007. *A Climate Of Injustice: Global Inequality, North–South Politics, and Climate Policy*. Cambridge, MA: MIT Press.

Robinson, N. A., ed. 1992. *Agenda 21 and UNCED Proceedings, Volumes 1 and 2*. New York: Oceana Publications.

Rossi, C. 1997. Controle da emissão de gases divide FHC e Clinton. *Folha de Sao Paulo*, October 10, A-15.

Sandbrook, R. 1997. UNGASS has run out of steam. *International Affairs*, **73**, 641–54.

Sebenius, J. K. 1991. Designing negotiations towards a new regime: the case of global warming. *International Security*, **15**(4), 110–48.

Sell, S. 1996. North–South environmental bargaining: ozone, climate change, and biodiversity. *Global Governance*, **2**(1), 97–118.

Shadlen, K. 2004. Patents and pills, power and procedure: the North–South politics of public health in the WTO. *Studies in Comparative International Development*, **39**(3), 76–108.

Sijm, J. P. M., Jansen, J. C., Battjes, J. J., Volkers, C. H. and Ybema, J. R. 2000. *The Multi-sector Convergence Approach of Burden Sharing: An Analysis of its Cost Implications*. Oslo: Center for International Climate and Environmental Research.

Sprinz, D. and Vaahtoranta, T. 1994. The interest-based explanation of international environmental policy. *International Organization*, **48**(1), 77–105.

Stavins, R. and Olmstead, S. M. 2006. An international policy architecture for the post-Kyoto era. *American Economic Review Papers and Proceedings*, **96**(2), 35–8.

Torvanger, A. and Ringius, L. 2002. Criteria for evaluation of burden-sharing rules in international climate policy. *International Environmental Agreements: Politics, Law and Economics*, **2**(3), 221–35.

Torvanger, A. and Godal, O. 2004. An evaluation of pre-Kyoto differentiation proposals for national greenhouse gas abatement targets. *International Environmental Agreements: Politics, Law and Economics*, **4**(1), 65–91.

UNFCCC (United Nations Framework Convention on Climate Change) 2008a. Ad-Hoc Working Group on Long-Term Cooperative Action (AWG-LCA). Submissions by Parties (Brazil, China, G-77). August 25, 2008.

UNFCCC 2008b. China's Views On Enabling The Full, Effective And Sustained Implementation Of The Convention Through Long-Term Cooperative Action Now, Up To And Beyond 2012. September 28, 2008. Available online: http://unfccc.int/files/kyoto_protocol/application/pdf/china_bap_280908.pdf.

Victor, D. 2001. *The Collapse of the Kyoto Protocol and the Struggle to Slow Global Warming*. Princeton, NJ: Princeton University Press.

Wade, R. 2003. What strategies are viable for developing countries today? The World Trade Organization and the shrinking of development space. *Review of International Political Economy*, **10**(4), 627–44.

Wapner, P. 1995. Politics beyond the state: environmental activism and world civic politics. *World Politics*, **47**(3), 311–40.

Wheeler, D. and Ummel, K. 2007. *Another Inconvenient Truth: A Carbon-Intensive South Faces Environmental Disaster No Matter What the North Does*. Working Paper Number 134, Center for Global Development, Washington, DC.

Woods, N. 1999. Good governance in international organizations. *Global Governance*, **5**(1), 36–61.

Ybema, J. R., Battjes, J. J., Jansen, J. C. and Ormel, F. 2000. *Burden Differentiation: GIIG Emissions, Undercurrents and Mitigation Costs*. Oslo, Norway: Center for International Climate and Environmental Research.

Young, O. R. 1994. *International Governance: Protecting The Environment In A Stateless Society*. Ithaca, NY: Cornell University Press.

5

Fair decision making in a new climate of risk

W. NEIL ADGER AND DONALD R. NELSON

Introduction

Adaptation to the impacts of climate change is happening now. No-one asked for the opportunity to adapt – clearly it is a set of actions in a situation caused by past and present human-induced change and hence a manifestation of past inequitable use of the earth's resources (Adger *et al.*, 2006). But adaptation is a central part of the changing landscape of human security and insecurity brought about by a changing climate.

Justice in the context of adapting to this new set of risks does not, therefore, simply involve a set of decisions on intertemporal and intergenerational equity involving trade-offs between unrepresented future agents and present actors and their interests. Rather, present day adaptation to climate change itself involves those justice dilemmas inherent in implementing sustainable and equitable development. The actions to maintain security are structured by present and historic inequalities in the distribution of access to resources and to power and decision making. There is now incontrovertible evidence of observed warming and trends in climate variability, and on the impacts of these trends on plants, animals, water, glaciers and ice is already apparent (Rosenzweig *et al.*, 2007). These impacts are consistent with projected future climate change, though the possibility of significant and even catastrophic impacts outside the assessed range is also real (Schellnhuber *et al.*, 2006). But the key point is that adaptation is already underway (Adger *et al.*, 2007). Hence, decisions to adapt to climate change to maintain human security are being taken in every part of the world and the equity implications of these decisions are becoming manifest and critical.

The thesis in this chapter is that one element of human security is the ability to shape one's own future and the agency to affect one's own resilience, both individually and collectively. Hence it follows that the issues of process and procedural fairness are central to human security. Human security is enhanced through access to

Climate Change, Ethics and Human Security, eds. Karen O'Brien, Asunción Lera St.Clair and Berit Kristoffersen. Published by Cambridge University Press. © Cambridge University Press 2010.

decision making and the empowering effect of participation in planning for the future under new climate risks as well as through material outcomes and consequences of those decisions. This is apparent both in situations where people recover from or protect themselves from weather-related events, and in everyday management of climate-sensitive resources on which people depend for their life and livelihoods (Morton, 2007).

How are decisions about adaptation made? In this chapter, we document decisions on resource management in the face of climate change or changing climate-related stress and show how institutions have been challenged in facing risks. We examine how issues of uncertainty, risk, and representativeness are central to the sustainability and legitimacy of present day governance and how climate change potentially moves governance away from equitable adaptation. The chapter draws on work on the management of coastal and marine resources in the Caribbean and water resources planning by newly evolving water resource committees in Brazil to identify the equity of process in adaptation decision making.

What are the implications of these observations? First, we argue that adaptation to climate change will face huge challenges and barriers to implementation given incommensurate values among competing interests. Second, we argue that there is therefore, a significant role for public collective action in adaptation. These roles include, first and foremost, the identification and protection of vulnerable populations from increased harm due to climate change. And more fundamentally, the issues raised point to the need for diversification of legitimate knowledge and interests in processes of adaptation. In short, structures for much environmental governance are likely to face major challenges associated with global changes that we have only yet begun to glimpse.

Adaptation to imposed climate change: vulnerability and access to decisions

The impacts of global climate change are already manifest and are being experienced in diverse ways in many different parts of the world. The IPCC Fourth Assessment Report (in its Impacts, Adaptation and Vulnerability Working Group) (IPCC, 2007) shows that there are no winners from the impacts of climate change. No country is immune. The losers, in terms of human health, misery, relocation and uncertainty, are not compensated for by some marginal and as yet unrealised increased productivity in forests, fish or specific yields in agricultural crops in specific places in the temperate regions of the world. The debates on where the impacts will fall often focus on the differential effects of climate change on countries, because the impacts of climate change are typically presented and projected at the global, continental or national levels. Yet localities and communities face differential climate impacts and have different vulnerabilities. At the same time, national governments do not necessarily forward the interests of their citizens equally or fairly.

But the IPCC Report also provides stark evidence of how people are being affected and who is most vulnerable to expected impacts in the future. Poor and marginalised people are most at risk from flooding, rising sea levels and wild weather. This is true for Europe and for the United States, as the Hurricane Katrina experience shows. The vulnerability of poor and marginalised people and places is all the more urgent for developing countries. But from our assessment of evidence on adaptive capacity within the IPCC, it is clear that the capacity to adapt is uneven both across and within all societies (Adger *et al.*, 2007). Overall, the IPCC, for example, concludes that young children and elderly are vulnerable in every country, particularly to heatwaves and extreme weather.

The vulnerable find it difficult to adapt for two reasons. First, they tend to be highly exposed, highly sensitive and have low adaptive capacity – this is how vulnerability is defined in analysis of climate change. There are many examples where low adaptive capacity translates into inability to adapt, sometimes with significant or disastrous consequences. Elderly people inherently vulnerable to extreme weather are the first to experience high mortality during heat waves and cold spells, as shown in the European heatwave in 2003 (Poumadère *et al.*, 2005).

The second reason that the vulnerable find it difficult to adapt is that they have limited access to public assistance and external resources. In other words, the capacity to adapt to climate change impacts is intimately bound up with capacity to access decision makers, engage in political processes, and access resources at times of crisis.

Adaptation therefore involves, we argue, issues of fair process as well as fair outcomes. Without fair decision-making institutions, fair outcomes will only ever be coincidental. Indeed some philosophical positions insist that fair process is predominant and the only necessary condition for justice. Fair process in democratic structures, as opposed to fair process in market exchange, is certainly important for the legitimacy of decisions concerning outcomes.

Fair decision making, as we use the term here, concerns how and by whom decisions on adaptive responses are made, and the recognition and participation of individual voices, and ultimately the legitimacy of the decisions. In discussing procedural equity we refer to fairness in access to decision-making institutions, which pertains to individuals, groups or nations, in line with those theories related to democratic decision making (Young, 1990). The issue of where justice should lie is contested – some theories stress differences among individual citizens that need to be addressed in fair process, while others argue for collective and group represent-ation (Young, 1990, 2000). The central issue here is that procedural justice deals with what Iris Young (1990: 23) calls 'democratic decision-making procedures as an element and condition of social justice.' Thus procedural justice does not refer narrowly, in this instance, to the procedures for the allocation of rights to property and goods. Procedural justice is interpreted through fairness in rules concerning resources to deal with decision-making elements such as voice, recognition and representation.

Adaptation to climate change is made up of uncoordinated choices and actions of individuals, firms and organizations in the face of present and expected climate change impacts. But it also involves collective action and choice at local, national and international levels, as well as cross-scale interactions. Adaptation is constrained by antecedent decisions and the existing institutional framework that engenders a particular distribution of resources, wealth and power. Collective choices bring up issues such as representation, participation, procedure and assent that do not characterise individual choices.

The governance of adaptation: inclusion and legitimacy

Considering the variety of future climate projections, there is a large pool of uncertainty regarding what our earth will be like in the mid- to long-term future. Even when projections correspond, it is not always clear how the climatic changes will translate into the physical environment. What is clear, however, is that the earth will be strikingly different from what it is today. The climate and the array of earth's environments have never been static. However, the rate and magnitude of change are beginning to approximate scales to which humans and societies are sensitive (Overpeck, 2006). Whether the projected outcomes of change are optimistic or, as is sometimes the case, apocalyptic, the changing climate will offer 'windows' of opportunity for taking purposeful adaptation decisions. Adaptation, as we have mentioned, however, is not necessarily an egalitarian process. If past experience is any guide, even as society adapts it tends to replicate the social and institutional structures that previously defined vulnerable groups and populations (Glantz and Jamieson, 2000; Wisner, 2001; O'Brien 2006). Nevertheless, climate change presents opportunities; not only in response to current events but in preparation for projected changes. It presents a clear entry point for rethinking the processes through which societies make adaptation decisions.

Adaptation is not simply about reducing risk. It is about creating the conditions for response, not only within a central structure but also throughout society as a whole. Ensuring that response capacity widely provides the flexibility required to address uncertain and unpredictable changes. For a number of years decentralised, deliberative bodies have been increasingly promoted as a way to help engage the participation of all citizens in decision making such creating bodies, it is argued, that are more flexible and responsive to local needs. The composition and attributes of these bodies differ between contexts but the basic premises are the same. First, it must be accepted that the policy goal involves some common good (Brunner *et al.*, 2005). The bodies should also encompass the ideals of inclusiveness, representativeness, procedural fairness, deliberativeness, publicity, equality, transparency and legitimacy, although different arrays of characteristics are apparent in different situations (Christiano, 1996; Young, 2000; Leach, 2006). Since climate change adaptation involves collective decision making at diverse scales, from the UNFCCC through to cooperation between farmers on their irrigation, modes of governance are diverse as well as interconnected (polycentric and multi-level in the political science terminology). There is no idealised situation or blueprint. And indeed, the ideal has been critiqued by a pragmatic political economy perspective that suggests that failure to recognise and address power imbalances between participants simply reinforces existing inequalities and the perpetuation of narrow interests (Adger *et al.*, 2005; Plummer and Armitage, 2007).

Despite these shortcomings, bringing together an array of actors on common footing offers many potential benefits. The nature of environmental change is highly complex as a result of scale, uncertainty and competing knowledge systems, combined with significant social inequality and vulnerabilities. There are numerous actors, with divergent and often incommensurable values. Approaching these issues through deliberative, co-management schemes provides pathways to overcoming management difficulties while offering non-instrumental benefits as well. There is the potential of enhanced efficiency of decision making, increased trust in government, decisions that are appropriate to the local social and ecological context, and increased capacity at the local scale to undertake, for example, monitoring and enforcement (Brosius *et al.*, 1998; Brown *et al.*, 2002; Dolšak and Ostrom, 2003; Brunner *et al.*, 2005). In addition, greater participation gives voice to vulnerable and marginalised stakeholders, recognition of diverse needs and knowledge systems, and an increase in the depth of civil society and citizenship (Dryzek, 2000; O'Neill, 2001).

In its most idealised form, co-management provides the space for internal learning: both institutional and ecological learning. Through provision of this space, adaptive governance strategies recognise the need for flexibility in light of uncertain futures and the need for continual improvement. Decisions are considered as one step in an iterative process, and actors use mistakes as a basis for re-evaluating and reorienting goals and strategies as parameters and knowledge change. This type of learning-oriented governance regime as applied in adaptive management strategies is dependent on continuously updated information in order to make evaluations. Such information could come from traditional science, but also from local knowledge systems that provide insights into functioning of local ecosystems and their linkages with the social system (Gadgil *et al.*, 2003; Brunner *et al.*, 2005). The process recognises the inherent value in participatory procedures. Thus, in addition to providing better management of social–ecological systems, adaptive management captures the ideals of the deliberative decision-making process including equity and inclusiveness.

Adaptation to our changing climate is already happening. The increased focus on projected change and the fact that the information is forward looking (rather than evaluation of a past event) provides unique opportunities for the ways in which adaptation decisions are made. The challenges are, however, significant. Below, we discuss two cases of resource management strategies designed to contend with variable and changing climates. Both management strategies contain elements of devolution, deliberation and learning. We highlight the successes and difficulties in each of these cases as a starting point to reflect on the challenges to adaptation processes in the future.

Neck deep in water governance in Brazil

In response to growing civil pressure and following on the heels of several similar state-level initiatives, the Brazilian Federal Government adopted a new water resources management system in 1997. The principles of the new system included integrating sectoral policies, devolving watershed management to local river basin committees, and ensuring the active participation of the range of stakeholders in each management area (Imprensa Nacional, 1997). Federal legislation provided the skeleton of the new management system at the local level. The states are responsible for fleshing out the details of how the legislation is implemented. Since 1997 all states have formulated water laws that meet the federal requirements and which specify the design and implementation of formal institutions and relationships (Formiga-Johnsson, 2005). Although these laws incorporate the spirit of devolved and deliberative decision making, results have not always lived up to the idealised expectations.

The first point, although seemingly obvious, is relevant to fairness of processes. The passage of legislation does not translate directly into practice. The federal laws mandated the decentralisation of water management power. But as Abers (2007) notes, this process encountered two problems. Power was supposed to be devolved to actors at the river basin management level, but neither political nor administrative power coincided with the physical territory of basins and new institutions had to be created. In addition, the water laws mandated an integrated management approach, contrary to the highly segmented approach in place at the time. This mandate essentially created a new policy field. As a result of these two mandates the federal government essentially 'granted power to a system yet to be constructed' (Abers and Keck, 2006: 617). The development of the necessary management systems and an integrated management approach are coming together thorough processes of trial and error, and none of the systems have been perfected.

As one would expect, each state interpreted the ideals of civil society, participation, inclusion and deliberation in different ways. São Paulo, for example, required one-third of all committee members to be representatives of civil society organisations, which include associations of water users, research organisations and organisations specialised in water and socioeconomic issues. In Paraná, on the other hand, only one-fifth of the committee must represent civil society, which, in addition to including organisations similar to those in São Paulo, also includes municipal representatives (Brannstrom *et al.*, 2004). In other regions there are cases of committees that have grown from the grassroots. These organisations, such as Manuelzão in Minas Gerais, started independently of legislation in response to the demands of local populations and have organically merged into the space created by the federal water laws (Abers, 2007).

The benefits that organisations such as Manuelzão brings to the communities go beyond providing space for an expanded civil society. First, it works to translate technical issues into ordinary language that is accessible to most of the residents (Abers, 2007). Accessibility of information is an important factor in maintaining equality in participatory bodies. Obscure knowledge can isolate individuals and be used to pursue the agenda of narrow interests (Lemos and Agrawal, 2006). The grassroots group also works to bring local knowledge and understanding of the issues into discussions within the committee. In addition to contributing to more efficient management of the resource, this provides the opportunity for the community to help define the boundaries of discussion, which is a principle of truly participatory deliberation. Rather than a focus on flood management and infrastructure the committee now is developing a vision for the basin as a whole (Frank, 2003).

In a recent survey, all 18 participating watersheds, located throughout the country, were experiencing climate-related stresses (Formiga-Johnsson and Lopes, 2003). These are exacerbated by growing populations, with increasing water demands and expanding urban areas. Recent climate projections for Brazil suggest significant changes over the next century (Marengo, 2007). Desiccation is projected for some already semi-arid regions, increased rainfall and extreme events are predicted in others. Brazil is a large country with a variety of climates and social–ecological contextual factors. Appropriate institutional and governance solutions will also vary accordingly. The flexibility built into the system at the federal level provides space for developing locally appropriate responses, yet it is a slow and often painful process.

In terms of inclusion and legitimacy of adaptation practices, two points pertain from this analysis of decentralised water management in Brazil. First, governance systems can change. The system is now more inclusive and takes a more holistic approach than a decade ago, bringing in issues of sustainability and wellbeing, and seeking behavioural as well as engineering and infrastructure solutions. Second, the issue of representation remains vexed. In this case the 'representation' of non-traditional stakeholders on committees (20% here, 30% there) is set by rules and actors external to those directly affected. These constraints on who is heard and included can, in effect, constrain the governance of adaptation to representation without power.

Swimming with the sharks in Tobago

In Tobago the social and ecological resilience of the Buccoo Reef ecosystem and the associated communities is faced with ongoing stresses associated with pollution loading, multiple uses for tourism and fishing, and the slow creeping problem of climate change (see Brown *et al.*, 2002). Although a Marine Protected Area, the Buccoo Reef Marine Park, was established in the early 1990s, it was ineffective. There was very little public support for the Park and poor enforcement of the regulations. The regulators saw conflicts between local marine resource users as a critical part of the problem, and perceived that self-interest was driving local people to exploit resources in an unsustainable manner. The system was both brittle and degraded, with apparently few incentives for effective collective action to support Marine Park management.

When we started to investigate local stakeholders' perceptions of the social ecological system and the Marine Park in particular, we developed a novel approach called 'Trade-off Analysis' which used future scenarios of local development as one means to explore how conflicts between users could be resolved and supported management strategies derived (Brown *et al.*, 2002). Interestingly, what emerged from this analysis was not a polarised or conflicting view, but, in fact, a large degree of agreement over what kind of future stakeholders wanted for this part of southwest Tobago. Out of this process came a new deliberative and inclusive planning regime, which we documented in Brown *et al.* (2002). The self-created group immediately solidified the informal interactions between individual agents and on this base grew the possibility of developing a more formalised co-management arrangement with the government decision makers.

The evolution of co-management arrangements brought about two critical changes at the community and government level. First, the various groups of previously conflicting stakeholders were mobilised to take both conservation and development actions together, as they recognised that they had more power as a group than as individuals. The second critical change arose as the multi-stakeholder group also realised that by acting collectively and agreeing on a single coherent message, they had greater influence with government agencies.

We suggest that in this case inclusionary and integrated learning-based coastal management contributes to the capacity to adapt to climate change in two ways. First, expanded networks act as a resource in coping with weather extremes. Thus, setting up inclusive governance structures means that these same people also work together on wider sustainable development issues. Second, informal institutions such as emerged to manage Buccoo Reef Marine Park are better at incorporating diverse knowledge and the learning that occurs in groups.

The positive aspects of this case of co-management are only one side of the story. At the same time, asymmetries in power between local resource users and government agencies meant that when the government agencies felt they were losing influence by the direction of some decisions, they could quickly re-establish dominance through imposing regulations. Government agencies tend to have more resources to engage in such linkages and hence to benefit from them. Thus the initial distribution of linkages may indeed skew the power relations between groups.

This case highlights that different social actors not only have differential adaptive capacity that affects their resilience, but they construct their knowledge of the social–ecological system and its desired state and the causes of change differently. While the inclusive management structures were much less formal than in the Brazil case, and in some ways more representative, the whole governance system was in fact fairly fragile and dependent on government support. This support was not forthcoming when the new inclusiveness threatened the status quo.

Conclusions

Adaptation to climate change will face huge challenges and barriers to implementation. These occur not least because of inequities in underlying social structures and antecedent decisions, and because of divergent and often incommensurate values among competing interests. Natural resources, such as water, forests and fisheries are governed through a mix of state, private and collective action regimes. Many of these management systems are, in effect, failing to halt over-exploitation or degradation of resources around the world for a diversity of reasons mainly outside the control of local stakeholders. We have attempted to show in this chapter that decentred democratic structures can potentially be effective in resource management and can even be empowering for previously excluded or vulnerable populations. The path towards adaptation requires radical changes to the status quo.

We argue that there is, therefore, a significant role for public collective action in adaptation in line with all serious examinations of the adaptation dilemma. The Stern Review, for example, suggests: 'in many cases, market forces are unlikely to lead to efficient adaptation' (Stern, 2007: 466). The roles for intervention to promote equitable and legitimate adaptation include, first and foremost, the identification and protection of vulnerable populations from increased harm due to climate change. But they also include the provision of accessible public good information on risks and uncertainty, and the protection of pure public goods such as cultural and natural heritage.

Neither of our examples of governance regimes and decision-making structures for resource management in the section above are direct examples of action to adapt to climate change per se. But they reveal some lessons concerning issues of representativeness and legitimacy that are generic and common to governance regimes for adaptation. The first steps in inclusionary and deliberative planning can make a significant difference to the legitimacy and fairness of adaptation decisions. In both cases highlighted this happens at a localised level. But Saleem ul Huq and Mizan Khan (2006) have demonstrated that these principles of inclusion are important even at national planning consultations, such as under the National Adaptation Programs of Action (NAPA) process. On the other hand, there are significant constraints on inclusive and adaptive management. Some underlying structures of ownership and control of resources represent strong inertia in implementing adaptive actions and promoting flexibility and resilience. So while deliberative planning and wider inclusion of stakeholders is much more common in many areas of resource management, Bronwyn Hayward (2008) argues that it has 'run into a cul-de-sac' whereby deliberation becomes simply a mechanism where some groups are able to exercise stakeholder rights over natural resources more effectively. Hence, deliberative planning and adaptive governance in general:

far from including oppressed voices, [deliberative governance] becomes a time bound as well as place bound exercise of private property rights and the transformative and critical potential of communicative of deliberative democracy is lost from view.

(Hayward, 2008: 83)

In summary, at present, structures for the governance of adaptation and managing the risks of climate change often lack legitimacy and meaningful representation. They are only likely to be further stretched given the resource scarcity and potential conflicts caused by climate change impacts.

A key unresolved question is whether adaptive learning is more likely to take place under stress or in anticipation of risks. When the impacts of climate change are already affecting people's lives and livelihoods, their options are often narrowed. There is an assumption in much of the science of climate change that adaptation will be stimulated by extreme events towards more sustainable pathways of development. But the experience of Hurricane Katrina in the United States shows that events can paralyse as much as stimulate action towards enhancing human security. We argue here that ensuring appropriate decision-making structures is the best way to ensure resilience and learning in the face of new and unforeseen risks.

References

Abers, R. N. 2007. Organizing for governance: building collaboration in Brazilian river basins. *World Development*, **25**, 1450–63.

Abers, R. N. and Keck, M. 2006. Muddy waters: the political construction of deliberative river basin governance in Brazil. *International Journal of Urban and Regional Research*, **30**, 601–22.

Adger, W. N., Brown, K. and Tompkins, E. 2005. The political economy of cross-scale interactions in natural resource management. *Ecology and Society*, **10**(2), 9. Available online: http://www.ecologyandsociety.org/vol10/iss2/art9/.

Adger, W. N., Paavola, J., Huq, S. and Mace, M. J., eds. 2006. *Fairness in Adapting to Climate Change*. Cambridge, MA: MIT Press.

Adger, W. N., Agrawala, S., Mirza, M. M. Q. *et al*. 2007. Assessment of adaptation practices, options, constraints and capacity. In M. L. Parry, O. F. Canziani, J. P. Palutikof, C. E. Hanson, and P. J. van der Linden, eds., *Climate Change 2007: Impacts, Adaptation and Vulnerability. Contribution of Working Group II to the Fourth Assessment Report of the Intergovernmental Panel on Climate Change*. Cambridge: Cambridge University Press, pp. 719–43.

Brannstrom, C., Clarke, J. and Newport, M. 2004. Civil society participation in the decentralisation of Brazil's water resources: assessing participation in three states. *Singapore Journal of Tropical Geography*, **25**(3), 304–321.

Brosius, J. P., Tsing, A. L. and Zerner, C. 1998. Representing communities: histories and politics of community-based natural resource management. *Society and Natural Resources*, **11**, 417–34.

Brown, K., Tompkins, E. L. and Adger, W. N. 2002. *Making Waves: Integrating Coastal Conservation and Development*. London: Earthscan.

Brunner, R. D., Steelman, T. M., Coe-Juell, L. et al., eds. 2005. *Adaptive Governance: Integrating Science, Policy, and Decision Making*. New York, NY: Columbia University Press.

Christiano, T. 1996. *The Rule of the Many: Fundamental Issues in Democractic Theory*. Boulder, CO: Westview Press.

Dolšak, N. and Ostrom E. 2003. The challenge of the commons. In N. Dolšak and E. Ostrom, eds., *The Commons in the New Millennium*. Cambridge, MA: MIT Press, pp. 3–34.

Dryzek, J. S. 2000. *Deliberative Democracy and Beyond: Liberals, Critics, Contestations*. Oxford, UK: Oxford University Press.

Formiga-Johnsson, R. M. 2005. *Institutional and Policy Analysis of River Basin Management: The Alto-Tietê River Basin, São Paulo, Brazil*. World Bank Policy Research Working Paper 3650. Washington, DC: World Bank.

Formiga-Johnsson, R. M. and Lopes, P. D., eds. 2003. *Retratos 3 x 4 das Bacias Pesquisadas*. Brasília: FINATEC.

Frank, B. 2003. Uma historia das enchentes e seus ensinamentos, In B. Frank and A. Pinheiro, eds., *Enchentes na Bacia do Itajai: 20 Anos de Experiências*. Blumenau: IPA and Edifurb, pp. 15–62.

Gadgil, M., Olsoon, P., Berkes, F. and Folke, C. 2003. Exploring the role of local ecological knowledge in ecosystem management: three case studies. In F. Berkes, J. Colding and C. Folke, eds., *Navigating Social-Ecological Systems: Building Resilience for Complexity and Change*. Cambridge, UK: Cambridge University Press, pp. 189–209.

Glantz, M. and Jamieson, D. 2000. Societal response to Hurricane Mitch and intra- versus intergenerational equity issues: whose norms should apply? *Risk Analysis*, **20**, 869–82.

Hayward, B. 2008. Let's talk about the weather: decentering environmental deliberation. A response to Iris Young. *Hypatia: Journal of Feminist Philosophy*, **23**, 79–98.

Huq, S. and Khan, M. 2006. Equity in National Adaptation Programs of Action (NAPAs): the case of Bangladesh', in W. N. Adger, J. Paavola, S. Huq and M. J. Mace, eds., *Fairness in Adapting to Climate Change*. Cambridge, MA: MIT Press, pp. 181–200.

Imprensa Nacional 1997. Lei no. 9.433. Diário da União.

IPCC (Intergovernmental Panel on Climate Change) 2007. *Climate Change 2007: Impacts, Adaptation and Vulnerability. Working Group II Contribution to the Intergovernmental Panel on Climate Change Fourth Assessment Report Summary for Policy Makers.* Geneva: World Meteorological Organization. Available online: www.ipcc.ch.

Leach, W. D. 2006. Collaborative public management and democracy: evidence from Western watershed partnerships. *Public Administration Review*, **66**, 100–110.

Lemos, M. C. and Agrawal, A. 2006. Environmental governance. *Annual Review of Environment and Resources*, **31**, 297–325.

Marengo, J. A. 2007. Caracterização do clima no Século XX e cenários no Brasil e na América do Sul para o Século XXI derivados dos modelos de clima do IPCC. São Paulo: Ministério do Meio Ambiente, Secretaria de Biodiversidade e Florestas, Diretoria de Conservação da Biodiversidade.

Morton, J. F. 2007. The impact of climate change on smallholder and subsistence agriculture. *Proceedings of the National Academy of Sciences*, **104**, 19680–5.

O'Brien, K. 2006. Are we missing the point? Global environmental change as an issue of human security. *Global Environmental Change*, **16**, 1–3.

O'Neill, J. 2001. Representing people, representing nature, representing the world. *Environment and Planning C: Government and Policy*, **19**, 483–500.

Overpeck, J. T. 2006. Abrupt change in earth's climate system. *Annual Review of Environment and Resources*, **31**, 1–31.

Plummer, R. and Armitage, D. 2007. A resilience-based framework for evaluating adaptive co-management: linking ecology, economics and society in a complex world. *Ecological Economics*, **61**, 62–74.

Poumadère, M., Mays, C., Le Mer, S. and Blong, R. 2005. The 2003 heat wave in France: dangerous climate change here and now. *Risk Analysis*, **25**, 1483–94.

Rosenzweig, C., Casassa, G., Karoly, D. J. *et al.* 2007. Assessment of observed changes and responses in natural and managed systems. In M. L. Parry, *et al.*, eds., *Climate Change 2007: Impacts, Adaptation and Vulnerability. Contribution of Working Group II to the Fourth Assessment Report of the Intergovernmental Panel on Climate Change.* Cambridge, UK: Cambridge University Press, pp. 81–131.

Schellnhuber, H. J., Cramer, W., Nakicenovic, N., Wigley, T. and Yohe, G., eds. 2006. *Avoiding Dangerous Climate Change*. Cambridge, UK: Cambridge University Press.

Stern, N. 2007. *Economics of Climate Change: The Stern Review*. Cambridge, UK: Cambridge University Press.

Wisner, B. 2001. Risk and the neo-liberal state: why post-Mitch lessons didn't reduce El Salvador's earthquake losses. *Disasters*, **25**, 251–68.

Young, I. M. 1990. *Justice and the Politics of Difference*. Princeton, NJ: Princeton University Press.

Young, I. M. 2000. *Inclusion and Democracy*. Oxford, UK: Oxford University Press.

Part III

Ethics

6

Ethics, politics, economics and the global environment

DESMOND McNEILL

The 'market' is a bad master, but can be a good servant.
Chakravarty (1993: 420)

Introduction

The above quotation offers a broad comment on the merits of a mixed economy and it effectively sums up the main argument in this chapter – namely that market instruments can and should be used by governments to confront global environmental challenges, and more specifically climate change, but that the market needs to be guided, rather than given free rein. Economic instruments that depend on market forces, including taxes and subsidies, are indeed very powerful; and, in the right hands, they have the potential to achieve the massive behavioural changes that are required to meet the challenges of climate change. However, to allow the market alone to determine how resources are to be allocated is not simply to risk inequitable and inefficient outcomes; it is to abrogate moral responsibility. Faced with a challenge as enormous as climate change, it can be considered reprehensible for a government (or numerous governments acting in concert) simply to say 'let the market decide'. At risk is not only the human security of the populations of the countries concerned, but people of all nations, in the present and, to a far greater extent, in the future.

The challenge of climate change can be summarised as follows. Global warming imposes high costs on present generations, especially the poor, and will impose even higher costs on future generations. It is widely agreed that something needs to be done to remedy this, but there is a lack of collective will, combined with disagreement as to what sort of measures are appropriate. These two problems are closely linked: as long as it is unclear how to act, it is even more difficult to gain agreement that drastic action is necessary. The situation is particularly challenging because the threat is supranational. It is not enough that one or a few countries apply the

Climate Change, Ethics and Human Security, eds. Karen O'Brien, Asunción Lera St.Clair and Berit Kristoffersen. Published by Cambridge University Press. © Cambridge University Press 2010.

necessary economic instruments; all the major economies of the world need to do it. Paradoxically, one of the hindrances to such concerted action could be existing international agreements, such as those under the World Trade Organisation, whose stated purpose is to constrain national policies that 'distort' market forces.

This situation poses a very serious challenge in terms of ethics, politics and economics. It is an ethical challenge insofar as present generations, and particularly the wealthy in high-energy, resource-consuming societies, have a moral obligation to address the problem (see also Chapters 2, 7, 8 and 10). It is a political challenge because without political will and appropriate governance structures, no solution will be found. It is an economic challenge because market instruments, of which economists have expert knowledge, have the capacity to make a major contribution to solving the problem – but are unlikely to do so as long as traditional economic approaches are followed. Economics is a discipline which has great powers of analysis and is readily linked to effective instruments of policy; but it needs to be the servant and not the master of policy makers, and subservient, not oblivious, to ethical considerations. In this chapter I seek to show how standard economic analysis frames issues such as climate change in a way that excludes ethical and political dimensions, and tends rather to conceal the social inequity and environmental costs that climate change brings. This is particularly problematic in view of the dominant position that economics enjoys, in relation to other social sciences, when it comes to advising policy makers.

My objective, in brief, is to highlight the need to link ethical, political and economic considerations when addressing climate change – focusing especially on the economic. I begin with the ethical dimension, adopting a very pragmatic, descriptive and non-normative approach. Rather than prescribing how people ought to behave, I take as my starting point how the majority of people actually do behave in relation to the different ethical dimensions of sustainable development. Next, I consider the political challenge involved in relation to climate change, which requires collaboration on a global scale between at least the most powerful nations. (This relates also to the limitations of political science and political philosophy, as highlighted by Gardiner, Chapter 8). Next, I consider the role of the market, identifying and discussing its strengths and weaknesses, including the damaging effects that it can have on society. I consider how standard mainstream economics deals with questions of social welfare, and especially environmental issues, through the use of cost–benefit analysis (CBA). Although acknowledging that CBA goes some way to remedying the shortcomings of a purely market-based approach, I show how even this is flawed – inspired, as it is, by a rather uncritical faith in 'consumer preferences'. I apply this argument specifically to the issue of the discount rate and suggest that the (market-based) rate commonly favoured by economists does not truly reflect the consideration which people actually give to future generations. The argument presented may seem quite radical to economists, but to the layperson appear as no more than common sense. The recommendation that follows is that society should make use of market instruments in order to combat climate change, but should not unquestioningly adopt the market-based discount rate as the basis for decisions concerning the future of the planet.

The ethics of sustainable development in practice

Sustainable development is a huge and complex challenge, not least in ethical terms. The nature of this ethical challenge may be summarised in terms of the three types of obligation to which sustainable development relates: (1) to people who are already living; (2) to people who are not yet born; and (3) to species other than humans. Given three such different types of obligations, individuals and groups are often faced with a serious moral dilemma. To make the nature of this dilemma clear, I refer to the perhaps controversial – and certainly uncomfortable – concept of a moral gradient (McNeill, 2007).

The central idea of a moral gradient is that the extent of one's moral obligations to others is not absolute, but instead varies according to what can be considered one's 'moral distance' from others (McNeill, 2007). The further away the other is perceived to be, the less is the extent of one's obligation to them. 'Moral distance' refers to three different dimensions of distance; three different ways in which other beings may differ from oneself in ways that are morally significant. First, there is the distance between one person and other people alive at the same time – what might be called social distance. Second, there is the distance between one person and others who do not yet exist – what might be called temporal distance. Third is the distance between a human being and other species that inhabit the globe: animals, plants, etc. This might be called species distance. It seems to be widely held (although seldom explicitly stated) that the extent of one's moral obligation declines with moral distance along each of these dimensions. Thus, for example, a person's obligation to another human being may be considered greater than his/her obligation to a dog (a view criticised, for example, by Peter Singer, 1990).

The moral gradient is by no means a smooth gradient, varying uniformly with each of the three moral distances (McNeill, 2007). Discontinuities may appear in quite abrupt steps. In the case of social distance, for example, there are strong discontinuities at the boundary of the family and of the nation. Other discontinuities can reflect the various ways in which one classifies others – for example, by religion or by language – which may imply allegiance and hence obligation. Consequently, the notion of 'social distance' is both variable and context-specific. But this does not affect the main argument: that the extent of one's perceived obligation to others varies considerably according to how one classifies that person or species (McNeill, 2007).

As a metaphor, one could consider that each individual stands at the top of an unevenly stepped pyramid with three faces. The pyramid slopes downwards in each of three directions to a base, which represents those to whom one feels no moral obligation. To some people this image may be startling or even abhorrent, for two reasons. The first is the very idea that moral obligations are not absolute but variable; this suggests that one's moral obligation to a peasant in India is less than that to a cousin or even to an unknown compatriot.[1]

The second reason why this metaphor may be abhorrent relates to the added claim that there are three dimensions; for this means that it is in theory possible that a person may feel a greater obligation to a pet or to a neighbour's as-yet unborn great-grandchild, than to a farmer in Malawi. Yet whether we like it or not, this is most often the case, representing the way we act, although perhaps not always the way we claim to act (McNeill, 2007).

In order to provide some empirical justification for this pyramid analogy, and to elaborate the point further, one may refer to various fields in which ethics is put into practice, such as law, medicine and economics. For example, in medicine we find some rather sophisticated moral gradients. In deciding what experiments are permissible and under what circumstances, the question arises: do we owe the same obligations to animals as we do to humans? Boards of medical ethics have been established to judge such issues and to lay down guidelines regarding the status of different animal species. This has been done in a clear and hierarchical way. The basic principle is that the more sensate the animal, the greater our obligation to be considerate of them. Here, then, we find that the steps of the moral pyramid are precisely codified. To take another example, in this case from the law, in the UK, there are six different categories of citizenship, so that the obligations of those that enjoy premier status towards the others vary according to clear and specified steps.

Whether or not one likes the idea of a moral gradient, there is ample evidence that people do in fact behave in a way that is consistent with such a concept. Yet it is a concept consistently argued against by exponents of global ethics or cosmopolitan views of social relations (see Dower, 1998; Gills, 2006). We might prefer to ignore this, and deny, for example, that we sometimes give greater weight to the unborn in our own country, or even to animals, than to those already born in a foreign country. Yet this empirical reality cannot be denied. Although we may readily agree that sustainable development is an important objective for us all, its moral implications are therefore more challenging than we perhaps wish to acknowledge (an issue also pointed out, and developed further, by Gardiner in Chapter 8). In summary, it is

---

[1] Such 'moral methodological nationalism' is abhorrent to Western philosophers. Also subject to criticism, by, for example, Thomas Pogge (2008) is the 'moral methodological territorialism', even of justice authors, such as John Rawls, who take their own Western social context as the defining normative point of departure for ethical analysis.

Table 6.1 *The moral gradient*

	Rich and poor	Humans and nature	Present and future generations
Discontinuous (not smooth)	– family – nation – race – religion...	– large mammals – small mammals – other animals – trees...	– children – grandchildren – not yet born ...
Varying over time*	Flatter in slope in recent decades	Flatter in slope in recent years?	Flatter in recent years? (now that a global threat exists)
Varying between cultures:	– North cf South: stronger bonds across nations? – South cf North: stronger bonds within the family?	– South cf North: greater respect for nature – in practice? Or only in theory?	North cf South: less (in practice)? Within South: very varied
Asymmetrical	Rich to poor	Humans to nature	Present to future

Source: McNeill, 2007 (slightly modified)
* There is variation over time in both North and South, but I here refer only to the former.

possible to distinguish three dimensions of rights and obligations: between rich and poor (currently living); between humans and other living beings; and between present and future generations. The 'moral gradient' along each of these dimensions is discontinuous (not smooth). As noted earlier, there may be sudden 'jumps' from one degree of obligation to another. Furthermore, the gradient can vary in slope over time, e.g. the moral views regarding duties towards foreigners today have changed significantly from a century ago, and they may change in the future as more people develop world-centric outlooks. The gradient may also vary in slope between cultures, e.g. the norms of Norwegians may be very different than those of, say, Tanzanians or Nepalese. Finally, the gradient is often asymmetrical: most would agree that the duties of the rich with regard to the poor are greater than the other way round. And in many cases, such obligations *cannot* be symmetrical: animals cannot have moral duties towards humans, nor future generations towards present generations.

The matrix in Table 6.1 summarises the differences between the three moral dimensions and the four characteristics. With regard to the issue of climate change, all three dimensions are significant. The 'nature' dimension has received rather less attention than the other two in the climate change literature – despite the long tradition of research on environmental ethics – reflecting the predominant anthropocentric tone of this debate. There has, however, been discussion as to the relative importance of the obligations to humans: should we be more concerned about the impacts of climate change on those already living in poor countries or on future generations? The issue is rendered more complex by the fact that the evidence suggests that those in tropical countries, where the preponderance of poor countries are located, are likely to be most adversely affected by climate change. This chapter, however, focuses mainly on the issue of future generations, since this is where I believe the policy prescriptions of mainstream 'market-based' economics most clearly diverge from the path many would advocate. The characteristics of this 'present/future generations' relationship, as summarised in the final column of the table, may be briefly spelled out.

Discontinuities are very marked; our feeling of obligation to (actually existing) children and grandchildren is hugely greater than to those who may, or may not, come into existence (and our feeling of obligation in relation to unspecified future generations or, for that matter, to future animals, is even less, since it is subject to 'double depreciation' – being attenuated across not one but two dimensions).

With regard to variation over time, I speculate that, in our own society, the gradient may have become flatter; that we have become more concerned for future generations than were our ancestors. If indeed this is the case, it is perhaps simply because it is only recently that we have had the power to substantially worsen the material conditions of future generations; we have become more concerned because we have a far greater power to do damage than was previously the case.

Regarding variation between cultures I am loath to generalise. It is easy to romanticise the practices of 'traditional societies'. No doubt some do indeed have attitudes to nature which are far more sustainable than our own in Western societies; but it is also the case that the mere fact of being poor limits the extent to which people are able to destroy the environment for future generations.

Last, but not least, the relationship between present and future generations is starkly asymmetrical. Not only can future generations (those not yet born) not speak for themselves, they cannot even bear silent witness. Indeed, some have used this fact to argue that our moral duties to future people are very limited – because they may or may not exist.

The politics of global environmental challenges

Many of the environmental challenges faced today are local, or at most national in scope, such as, for example, pollution. It is not unusual for the rich and powerful to reduce or avoid the costs imposed, either by insulating themselves against them or transferring the costs to another group. Examples of the former are where the dirt and noise of city streets is avoided by driving a private car or urban air pollution reduced by air-conditioning. Examples of the latter are where rubbish from prosperous parts of the city is dumped in poorer parts; or, on a national scale, where atomic waste is exported to poor countries. In some cases monetary compensation may be made (e.g. to the receiving country), but controversial political issues nevertheless arise concerning the distribution of costs and benefits between groups with varying degrees of power.

Environmental problems such as these affect people in all parts of the world. But there are some problems that are truly 'global' in scope – not merely in the sense that they are encountered in most places, but in the more narrow sense that they cannot simply be shifted elsewhere. Climate change is thus a global problem in a more fundamental sense than, for example, water scarcity and water pollution – even though these affect huge numbers of people. Nor can a particular country, or region within a country, entirely insulate itself against climate change. A protective dyke could, in theory, be built around a house, a city or even an entire country if it is technically possible (the Netherlands is spending many millions of Euros in building dykes) but rising water levels are only one outcome of climate change and the costs of this sort of adaptation are enormous. This is why radical mitigation measures are being debated and why climate change is such a challenge for 'global governance'. To reduce global warming will require effective global institutions and political will – at least among the largest economies. (Although Kyoto has been widely, and justifiably, criticised, it is worth recognising how radical even this small step has been, in terms of global governance.)

While the costs of adaptation can be charged to the individual – householder, city or country – the costs (and benefits) of mitigation will be shared, on terms not yet specified. The balance which each country chooses to adopt – between adaptation and mitigation – will vary; but not all are equally well placed to respond to the challenge of global warming. All are likely to suffer to some extent; but, broadly speaking, poor countries are likely to suffer more.

There are thus good grounds for adopting a normative stance, arguing that we (the rich) have a moral duty to do more than most people in order to reduce the extent of climate change; and that ethical values need to be changed. This argument is put forward by Gardiner, Caney and St.Clair in this volume, but my claim is more modest. I argue that the market-based discount rate, which economists take as the basis for giving advice, does not accurately reflect the values that people already have. I suggest that, if they had the opportunity to express their views, most people in rich countries (and perhaps also in poor) would be willing to sacrifice a very small part of their income in order to avoid the sort of catastrophic consequences that are likely to befall the world in the future if action to reduce global warming is not taken now. I also suggest, however, that instruments that operate through the market – taxes and subsidies – can make a very effective contribution to reducing global warming. The market is thus both the problem and the potential solution. To undertake concerted action at the international level will certainly be a major challenge; and political leadership will be required to connect widely agreed (ethical) ends with the available (economic) means. But much can be achieved, I suggest, even without changing people's values. This is not to say that people do not need to be better informed about the likely impacts of climate change. Quite the reverse, it is only when the majority of people come to understand the extent of this threat that they will be willing to translate their concern for the future of the world into support for governments that introduce the taxes and subsidies that are needed to transform energy systems and lifestyles to achieve dramatic reductions of greenhouse gas emissions. A continuous and effective information campaign may be necessary in order to ensure popular support for the use of these powerful economic instruments.

Economics and climate change

To better understand the strengths and limitations of economics, I will show how the rule of the market plays out in relation to the various issues summarised in Table 6.1 above. A good example is that of the market and the spatial gradient. Exporting pollution is an example of what occurs as a result of market forces. It was by drawing attention to this fact that the former Chief Economist at the World Bank, Lawrence Summers, incurred the wrath of environmentalists the world over. He wrote an internal memo which may be briefly summarised:[2]

'Dirty' Industries: Just between you and me, shouldn't the World Bank be encouraging MORE migration of the dirty industries to the LDCs [Less Developed Countries]? I can think of three reasons:

1. I think the economic logic behind dumping a load of toxic waste in the lowest wage country is impeccable and we should face up to that . . .
2. The costs of pollution are likely to be non-linear . . .
3. The demand for a clean environment for aesthetic and health reasons is likely to have very high income elasticity . . .

The problem with the arguments against all of these proposals for more pollution in LDCs (intrinsic rights to certain goods, moral reasons, social concerns, lack of adequate markets, etc.) could be turned around and used more or less effectively against every Bank proposal for liberalization

After the memo became public in February 1992, Brazil's then-Secretary of the Environment Jose Lutzenburger wrote back to Summers: 'Your reasoning is perfectly logical but totally insane . . . Your thoughts [provide] a concrete example of

[2] http://www.whirledbank.org/ourwords/summers.html

the unbelievable alienation, reductionist thinking, social ruthlessness and the arrogant ignorance of many conventional "economists" concerning the nature of the world we live in.'

The issue raises quite complex questions. A 'hard-line' market enthusiast might argue that exporting pollution was entirely defensible in ethical terms so long as both parties agreed to the trade. A less hard-line argument would be that it is defensible provided that all those affected by importing the pollution are adequately compensated (which would lead to a long debate about what is 'adequate', whether there is sufficient information as to the costs, whether this information is made available to those affected, etc). I do not need to analyse the case further; my point is simply that here the 'market' view led to a conclusion which many people regarded as ethically indefensible.

A second example concerns the market and the temporal gradient. The market heavily discounts future benefits and costs. In other words, it puts a lower value on costs and benefits which occur in the future: the further distant they are in time, the lesser their value (this is why, to oversimplify, one can earn interest – even allowing for inflation – on money deposited in the bank). Does this imply that we should adopt the market discount rate in determining the basis on which to value the costs and benefits that accrue to future generations? This has for many years been an issue in cost–benefit analysis, which is sometimes called social cost–benefit analysis, to emphasise the point that such analysis should take account not only of narrow financial revenues and expenditures but all costs and benefits that accrue, including those that are not reflected in market transactions. The challenge of global warming, which brings to the fore the issue of costs to future generations, has caused economists and others to focus once again, and rather more critically, on this issue.

The theoretical justification for discounting the future is more complex than simply referring to the market rate of interest, and includes factors such as risk, and the expectation that future generations will be richer (it is seldom considered that they may be poorer). In recent years, there has been a very active debate as to what discount rate should be used, with some proponents of sustainable development favouring the adoption of a much lower, or even zero, discount rate. This debate has been especially lively following the famous *Stern Review on the Economics of Climate Change*, the enormously influential 700-page report written by ex-World Bank Chief Economist Nicholas Stern for the British Government (Stern, 2006). I shall not try to summarise it, but focus immediately on the question of discounting and how Stern's approach has been received. Very many economists, and non-economists, have commented on the review – both favourably and unfavourably. For a representative, and authoritative, source of mainstream economic reaction the most appropriate reference is the *Journal of Economic Literature* which, in September 2007, published reviews by two leading experts, Nordhaus and Weitzman. Their views are very similar and deserve to be quoted at some length. I begin with Nordhaus, who notes that the Stern Review 'clearly and unambiguously' concludes that 'we need urgent, sharp, and immediate reductions in greenhouse gas emission' (Nordhaus, 2007: 701). However, he asserts:

The Review's radical revision of the economics of climate change does not arise from any new economics, science, or modelling. Rather, it depends decisively on the assumption of a near-zero time discount rate combined with a specific utility function.

(Nordhaus, 2007: 701)

In this he is quite right. But I would suggest that this is its merit: that the review is based on an ethical judgement about our responsibilities to future generations and on claims that this, not the market or 'positive' economic theory, should be our guide in taking the necessary steps. Nordhaus almost ridicules Stern:

The Review takes the lofty vantage point of the world social planner, perhaps stoking the dying embers of the British Empire, in determining the way the world should combat the dangers of global warming. The world, according to Government House utilitarianism,[3] should use the combination of time discounting and consumption elasticity that the Review's authors find persuasive from their ethical vantage point.

(Nordhaus, 2007: 691)

And he seems to claim that moral judgement has no more of a place in economics than it does in the natural sciences:

This approach does not make a case for the social desirability of the distribution of incomes over space or time of existing conditions, any more than a marine biologist makes a moral judgement on the equity of the eating habits of marine organisms in attempting to understand the effect of acidification on marine life.

(Nordhaus, 2007: 692)

Noting that 'The Review argues that fundamental ethics require intergenerational neutrality as represented by a near-zero time discount rate,' Nordhaus rightly points out that other approaches are possible, listing, for example, the principle 'that each generation should leave as much total societal capital . . . as it inherited' and other approaches that may be regarded as more radical than that of the Review. But he is wrong, I suggest, to say that each of these embodies 'Quite another ethical stance.' What they surely have in common is that they take greater account of the welfare of future generations than does the market. But it is the rule of the market that he appears to advocate, in concluding that 'because we live in an open-economy world of sometimes-competing and sometimes-cooperating relay teams (i.e. from

---

[3] A reference to Sen and Williams (1982: 16)

generation to generation), we must consider how the world's capital market will equilibrate.'

He concludes that 'The Review's unambiguous conclusions about the need for extreme immediate action will not survive the substitution of assumptions that are more consistent with today's marketplace real interest rates and savings rates.' He is quite right; but, as I shall argue below, the market-based discount rate does not adequately reflect people's preferences regarding future generations, which is why the marketplace should not be taken as our guide. I turn now to Weitzman, who demonstrates very clearly why the choice of discount rate is so crucial:

Global climate change unfolds over a time scale of centuries and, through the power of compound interest, what to do now is hugely sensitive to the discount rate that is postulated. In fact, it is not an exaggeration to say that the biggest uncertainty of all in the economics of climate change is the uncertainty about which interest rate to use for discounting ... This little secret is known to insiders in the economic of climate change, but it needs to be more widely appreciated by economists at large.[4]

(Weitzman, 2007: 705)

The theory is complex and need not be presented here. He proposes as a '"point guess-estimate" an annual rate of 2% each for discounting utility and discounting consumption, and the same figure for "a measure of aversion to interpersonal inequality and a measure of personal risk aversion."'[5] (Weitzman, 2007: 706). What the lay reader needs to know is that these three combine to give an aggregate figure of 6% per year. This contrasts with the Review's figure of 1.4%. Such is the power of compound interest that, as he points out: 'the present discounted value of a given global-warming loss from a century hence at the non-Stern annual interest rate of r = 6% is one *hundredth* of the present discounted value of the same loss at Stern's annual interest rate of 1.4%' (Weitzman, 2007: 708). In brief, on the basis of Stern's figures, the cost of global warming a century hence is 100 times greater than that calculated by the author using 'what most economists might think are decent parameter values' (Weitzman, 2007: 707).

Weitzman seems to be rather more willing than Nordhaus to recognise that the Review is not, and should not be, a purely economic document:

The Stern Review is a political document ... at least as much as it is an economic analysis and, in fairness, it needs ultimately to be judged by both standards. To its great credit, the Review supports very strongly the politically unpalatable idea ... that ... substantial carbon taxes must be levied.

(Weitzman, 2007: 723)

---

[4] And, I would argue, by non-economists also.
[5] These numbers are based, he claims, on 'tastes'; a popular term in economics that deserves more critical scrutiny.

His objection is that the Review 'predetermines the outcome' by adopting a very low discount rate, arguing instead for an intermediate discount rate of 2–4%. He refers to the Review's 'urgent tone of morality and alarm' and criticises it for not more openly revealing that its conclusions result from adopting 'discount rates that most mainstream economists would consider much too low' (Weitzman, 2007: 724).

What, then, do mainstream economists say about the discount rate? Economics occupies a privileged position among the social sciences in relation to policy makers; and the market occupies a privileged position within economics. This is not to say that all economists, or even all mainstream economists, are slavish advocates of the market. An extreme view, held by few, is that the market actually is 'perfect'. A more modest view, held by many, is that a 'market system' is, despite its many faults, better than any feasible alternative. A major practical advantage that the market enjoys is that it gives definite answers. Those who argue for a different basis for decision making and allocation of resources can be met with the claim that their favoured number is somehow 'arbitrary', unlike the value expressed by the market. (The recent financial crisis has helped to challenge this simplistic claim, since it has become apparent to everyone how difficult it is to establish the 'correct' market value of a bank.)

Economists are well aware of the limitations of the market; and social CBA has developed in recognition of this. Instead of the market, decisions are here based on what might be called a 'quasi-market': a market which more accurately reflects consumer preferences. Following this logic, a sophisticated field of study has developed and applied mainly in relation to the appraisal of projects in developing countries (Little and Mirrlees, 1969; Dasgupta *et al.*, 1970) and of environmental projects and policies (Pearce *et al.*, 1989; Pearce, 1998). Decisions based on CBA are certainly to be preferred to those based simply on the market, but a number of criticisms may nevertheless be made of this approach. These relate quite closely to the three dimensions discussed above and summarised in Table 6.1.

Let us begin with 'rich and poor'. The market, in effect, weights consumer preferences according to their purchasing power; and the willingness to pay of people who have high incomes is, of course, much higher than that of people with low incomes. It is true that CBA techniques can, in theory, be adapted to correct for this (by attaching higher weights to benefits and costs accruing to poor people), but this has rarely been done in practice. Considerable controversy was aroused in debate surrounding the contributions of economists to the work of the IPCC when critics reacted to the fact that the 'value' attributed to poor people's lives in the CBA calculations was far less than the 'value' attributed to rich people's lives (Pearce, 1995). Here we find one group – the poor – being inadequately represented.

More extreme than the case of the poor is the fact that other groups affected – animals and future generations – are not represented at all. Animals (and other elements that constitute 'nature') are not consumers; they have no voice. How one responds to this challenge depends on whether one is an adherent of an anthropocentric or non-anthropocentric view. In the former case, one may argue that this problem can be handled by methods already in common use in CBA, such as willingness to pay studies. In the latter case (those who believe that nature has an inherent value), one could in theory argue that the animal kingdom, or even 'nature' more generally, deserves to have its interests represented as if it were 'a consumer'. But it is hard to see what this might mean in practice, and a more likely response is simply to reject the 'consumer preference' view.

Similarly, future generations have no opportunity to express their voice through the market. A standard counter-argument is that we – present generations – can speak for them. But, as I shall argue below, the matter is not so simple. A still more fundamental criticism of valuations based on consumer preferences is that these are based on the preferences of *individual* consumers. It may well be the case that individuals, acting in concert, would express different values – but the market does not generally provide mechanisms for expressing such preferences. Some would go even further, and argue that the way in which the issue has been framed here is in itself misleading: the question, they might say, is not 'what is the value of nature?' or 'what is the value of future generations?', but rather 'what are we willing to sacrifice in order to ensure the continued existence of the world as we know it?'

My claim in this chapter is less radical, but constitutes, nevertheless, a major challenge to conventional economic wisdom. I here limit myself to the issue of future generations and the question of the discount rate. The importance of this issue has given rise to a burgeoning, and often rather technical, literature within economics (e.g. the work by Asheim *et al.*, 2001, on discounting and sustainability). This has been greatly stimulated by the controversial Stern Review discussed above. Most economists are in accord with Nordhaus and Weitzman, quoted above, albeit with some variation. One of the very few exceptions – among 'recognised' economists – is John Roemer, who is respected also as a political scientist and philosopher specialising in social justice. In brief, he rejects the standard position, as represented by, for example, Nordhaus, on two grounds:

[F]irst, the analyses almost always assume that the correct intergenerational ethic is utilitarianism, and second, the modification to discounted utilitarianism is often based on the ethical view that the decision problem for a society with many generations is ethically equivalent to the decision problem of an infinitely lived consumer.

(Roemer, 2009: 41)

Roemer's criticism of the standard economic approach is in many ways similar to my own, although his account is more narrowly and rigorously couched in the terms of welfare economics. Economists , he says, reveal their inadequate grasp of ethics by 'adopting the model of the infinitely lived consumer as ethically equivalent to a sequence of generations of human beings' (Roemer, 2009: 19). The alternative that he proposes draws upon the popular view that an appropriate intergenerational ethics should require *sustainability*. In formal terms, he expresses this as the adoption of a 'maximin' principle applied across generations: that 'the date at which a person is born should be viewed as arbitrary from the moral viewpoint' (Roemer, 2009: 22). As he notes, this is a 'distinctly anthropomorphic conception of sustainability' – weak rather than strong.

The implication he draws is that the discount rate that most analysts have adopted is far too small. The practical challenge, however, is how to determine what the rate should be. Here, Roemer makes a similar point to mine in relation to the market: criticising what might be called, following Whitehead, 'misplaced concreteness':

[M]any economists today use discounted-utilitarianism: not because it has a sound ethical foundation – at least for the discount rates commonly employed – but because it *gives a unique answer* to the problem. There is no good justification for this practice: it is an example of looking for the lost diamond ring under the street lamp, because that is the only place one can see! If (undiscounted) utilitarianism does not enable us to find the optimal policy in all problems, that means only that it is an incomplete ethical doctrine – not something to be ashamed of.

(Roemer, 2009: 19)

In brief, not only non-economists but also a few well-regarded economists challenge the assumption that the market provides the correct value to assign to the discount rate – the single, immensely powerful, figure that specifies the extent to which the well-being of future generations shall be taken into account in decisions taken by the present generation. As noted above, the stakes are very high: the discount rate favoured by Stern yields an estimate of future costs that is *100 times* the figure as calculated by mainstream economists. Scientific uncertainty regarding the rate of global warming and its effects on, for example, sea levels, thus pales into insignificance in comparison with uncertainty regarding the rate of discount. And the latter is not a scientific uncertainty that can be resolved simply by calling on the expertise of economists. This is for two reasons: first, this is not simply a technical but an ethical issue. Second, the issue is unprecedented; it arises because the world faces a challenge of a kind never encountered before. We therefore need to critically assess whether the methods we have used in the past, often with considerable success, may need to be radically revised in this new situation. The problem, and it is a significant problem, is how to establish an agreed figure to replace the discount rate, or, if necessary, to establish a new basis for decision making regarding decisions with very long-term consequences?

Conclusion

Economists are engaged by policy makers to assist them in taking decisions, but they tend to follow rather than lead public opinion: sometimes rather belatedly. Many economists make strong claims for the market, arguing that it is here that society expresses its values explicitly. The danger arises, however, that slavishly following the market may bring about results which run counter to the interests of both present and future generations.

In addition to being a very influential discipline, economics has powerful instruments of policy at its command – many of which are well suited to responding to the enormous challenge of climate change. Some of them impact on the behaviour of the consumer: for example, the imposition of a higher tax on fuel oil or provision of subsidies for roof insulation. Others impact on the behaviour of firms: for example, the provision of major financial incentives for investment in new technology or the imposition of higher effluent taxes. These are only some of the instruments available. Each of them involves interfering with – not giving free rein to – the market; and if used judiciously they can surely increase well-being.

It is for this reason that I argue that the market is part of the solution as well as part of the problem. And the same goes for economics. The study of economics can help us understand what is going on; and it can help us change what is going on – if, for example, we believe that current practice fails to take adequate account of the interests of future generations. That is why it is appropriate to conclude with a slightly modified version of the quotation with which this chapter began:

Economics is a bad master, but can be a good servant.

References

Asheim, G. B., Buchholz, W. and Tungodden, B. 2001. Justifying Sustainability. *Journal of Environmental Economics and Management*, **41**(3), 252–68.

Chakravarty, S.1993. Policy-making in a mixed economy: the Indian case. In *selected Economic Writings*. Oxford, UK: Oxford University Press.

Dasgupta, P., Marglin, S. and Sen, A. 1970. *UNIDO Guidelines for Project Evaluation*. New York, NY: United Nations.

Dower, N. 1998. *World Ethics: A New Agenda*. Edinburgh: Edinburgh University Press.

Gills, B. 2006. The global politics of justice.*Globalizations*, Special Issue, **3**, 95–8.

Little, I. and Mirrlees, J. 1969. *Manual of Industrial Project Analysis. Volume II*. Paris: OECD Development Centre.

McNeill, D. 2007. The ethics of sustainable development. In C. H. Grenholm and N. Kameragrauzis, eds., *Sustainable Development and Global Ethics*. Stockholm: Acta Universitatis Upsaliensis.

Nordhaus, W. 2007. A review of the Stern Review. *Journal of Economic Literature*, **XLV**, 687–702.

Pearce, D. W. 1998. *Economics and the Environment*. Cheltenham, UK: Edward Elgar.

Pearce, D. W., Markandya, A. and Barbier, E. 1989. *Blueprint for a Green Economy*.
    London: Earthscan.
Pearce, F. 1995. Global row over value of human life. *New Scientist*, 19 August 1995.
Pogge, T. 2008. *World Poverty and Human Rights*, 2nd edn. London: Polity Press.
Roemer, J. 2009. *The Ethics of Distribution in a Warming Climate*. Cowles Foundation
    Discussion Paper No. 1693. Yale University, New Haven, Connecticut.
Sen, A. and Williams, B., eds. 1982. *Utilitarianism and Beyond*. Cambridge and New York,
    NY: Cambridge University Press
Singer, P. 1990. *Animal Liberation*. New York, NY: New York Review of Books.
Stern, N. 2006. *Stern review on the economics of climate change*. London: Cabinet Office,
    HM Treasury.
Weitzman, M. 2007. A review of the Stern review. *Journal of Economic Literature*, **XLV**,
    703–24.

7

Human rights, climate change, and discounting

SIMON CANEY

In its Fourth Assessment Report, the Intergovernmental Panel of Climate Change (IPCC) has provided further confirmation that climate change will have serious impacts on human life. It reports that:

The global average surface temperature has increased, especially since about 1950. The updated 100-year trend (1906–2005) of 0.74°C ± 0.18°C is larger than the 100-year warming trend at the time of the TAR (1901–2000) of 0.6°C ± 0.2°C due to additional warm years. The total temperature increase from 1850–1899 to 2001–2005 is 0.76°C ± 0.19°C. The rate of warming averaged over the last 50 years (0.13°C ± 0.03°C per decade) is nearly twice that for the last 100 years.

(Solomon et al., 2007: 36)

In addition to this, the IPCC projects that temperatures will continue to rise. It employed six different Special Report of Emission Scenarios (SRES) and these all found that temperatures will rise by 2090–2099 as compared to the temperatures between 1980 and 1999. In some scenarios temperatures will increase by 1.8°C (the best estimate of the B1 scenario). In others temperatures will increase by 4.0°C (the best estimate of the A1FI scenario). If we examine the 'likely range,' then the lower limit is 1.1°C and the higher limit is 6.4°C (Solomon *et al.*, 2007: 70). Climate change will also involve a rise of sea levels. Again the IPCC employs six different SRES scenarios. In some, sea levels are projected to rise by 0.18–0.38 metres (B1 scenario) and on others the increase is projected to be 0.26–0.59 metres (A1FI scenario). It is crucial to note that these projections exclude "future rapid dynamical changes in ice flow" (Solomon *et al.*, 2007: 70). Furthermore, the situation looks to be getting worse rather than better. Carbon dioxide emissions stemming from fossil fuel use and industry have grown by more than 3% per annum during the 2000–2004 period (as compared to 1.1% per annum between 1990 and 1999) (Raupach *et al.*, 2007: 10 288).

These changes pose considerable ethical challenges. One concerns how we evaluate the impacts of climate change. For example, do persons have a right not

Climate Change, Ethics and Human Security, eds. Karen O'Brien, Asunción Lera St.Clair and Berit Kristoffersen. Published by Cambridge University Press. © Cambridge University Press 2010.

to suffer the ill effects of global climate change? Some have advocated a human right to a healthy environment and related concepts are affirmed in international law. One important starting point is the 1972 Stockholm Declaration of the United Nations Conference on the Human Environment. Principle 1 of the Declaration maintains that: "Man has the fundamental right to freedom, equality and adequate conditions of life, in an environment of a quality that permits a life of dignity and well-being, and he bears a solemn responsibility to protect and improve the environment for present and future generations."[1] More recently the United Nations Human Rights Council passed a resolution in 2008, stating that "climate change poses an immediate and far-reaching threat to people and communities around the world and has implications for the full enjoyment of human rights."[2] In this chapter, I wish to argue that persons do have a human right to a healthy environment (Caney, 2009b, 2009d; see also Nickel, 1993; Adger, 2004; Hayward, 2005). The current consumption of fossil fuels is, I argue, unjust because it undermines certain key rights. The remainder of the chapter maintains that this right should not be subject to a positive pure time discount rate.

The main argument

Anthropogenic climate change might be condemned on a number of grounds. I here argue that one, but only one, source of condemnation is that climate change jeopardizes human rights.[3] My argument is that employing the normal kinds of argument for justifying rights shows that persons have a right not to suffer from dangerous climate change. The argument can be stated relatively simply.

Step 1: First, as Joseph Raz (1986, Chapter 7, in general, and p. 166, in particular) has persuasively argued, to say that X has a right is to say that X has interests which are sufficiently weighty to impose obligations on others. This account can explain our use of the concept of rights. We hold that there are rights to activities such as freedom of expression and association and belief because these protect important interests which are weighty enough to impose obligations on others.

Step 2: Consider now climate change. This clearly jeopardizes several fundamental interests. Three in particular are worth stressing. First, climate change is likely to lead to widespread malnutrition and jeopardizes persons' interests in subsistence. According to recent estimates, a temperature increase of 2.5°C will result in an extra 45–55 million people suffering from hunger by the 2080s; a temperature increase of 3°C will result in an increase of 65–75 million people of

---

[1] See http://www.unep.org/Documents.multilingual/Default.asp?DocumentID=97&ArticleID=1503.
[2] This was agreed at the seventh session of the Human Rights Council on March 26, 2008 (A/HRC/7/L.21/Rev.1).
[3] For an important "non-rights" based analysis see Page (2006).

those who are threatened by hunger; and a temperature increase of 3–4°C will result in an increase of 80–125 million in that category (Hare, 2006: 179). Second, climate change undermines person's interests in being able to support themselves. Even if people are not subject to malnutrition their capacity to attain a decent standard of living is threatened by climate change. Rising sea levels, storm surges, and extreme weather events may destroy buildings and infrastructure, and thereby ruin businesses and industries. Decreased rainfall may obviously lead to crop failure. Some of those affected may not be threatened by malnutrition but their capacity to support themselves could nonetheless be greatly impaired. Third, climate change jeopardizes a fundamental interest in health. As many have noted, climate change can endanger human health through a number of different mechanisms, including (a) extreme weather phenomena, (b) heat stress, (c) vector-borne diseases (e.g., malaria and dengue), (d) water-borne diseases (e.g. cholera and diarrhoea), and (e) poor air quality (Patz *et al.*, 2000; McMichael *et al.*, 2003, 2004, especially pp. 1562–605; Kovats *et al.*, 2005; Confalonieri *et al.*, 2007). Climate change thus clearly damages key human interests.

Step 3: This, though, is not sufficient to establish that it violates people's rights because, as was noted above, persons have rights when they have an interest that is fundamental enough to impose obligations on others. The third step in my argument, then, is the interests cited above *are* indeed sufficient to impose obligations on others. Several considerations are relevant here. In the first place we should record that the interests in question are not trivial or frivolous interests – they are key to all persons. These interests are as fundamental as, say, other kinds of interests which we believe that should be protected by rights – interests such as the interest in freedom of speech or creed. Second, the evidence that is currently available suggests that averting dangerous climate change can be secured at a cost of several percent of global GDP per annum. For example, the Stern Review has suggested that it will only cost between -1% and 3.5% to stabilize the concentration of carbon dioxide in the atmosphere at 550pm, and it proposes 1% of GDP as the most likely cost (2007: 239). Similar estimates are made by Christian Azar and Stephen Schneider (2002). The most recent IPCC Assessment Report produces estimates that are higher but nonetheless entirely reasonable. For example, the "Summary for Policymakers" produced by Working Group III of the IPCC (those detailed to work on mitigation) concludes that the cost of stabilizing CO_2 concentration in the 445–535 range will involve a cut in the global GDP in 2030 of less than 3% (Barker *et al.*, 2007: 12). Were the costs excessive, then one might conclude that the interests in avoiding dangerous climate change are not fundamental enough to impose obligations on others. That an interest is vital is insufficient to generate a right to it until we know whether it is appropriate to hold others to be under a duty to it. However, the evidence just cited shows that the costs are reasonable.

On the basis of what we have seen so far we can say, then, that persons have fundamental interests in health, subsistence, and supporting themselves and that the duty to protect these interests from dangerous climate change is not unreasonably demanding on the appropriate would-be duty bearers. Given both of these considerations we may, then, conclude that the interests in question are sufficient to impose duties on others. As such, it follows that, on the Razian account affirmed earlier, people have a right not to suffer from climate change which jeopardizes these interests. Or put otherwise, they have rights to health, subsistence, and to be economically independent, and these rights are threatened by dangerous climate change.

Two further points should be made about this argument. First, note that if it is valid, it provides a defense of *human* rights because the interests cited are interests of all human beings. If the preceding argument is correct, all persons have this right not to be exposed to dangerous climate change. Second, note that the argument provides a different way of thinking about climate change. The prevailing intellectual framework employed to analyze climate change by policy elites is cost–benefit analysis. If my argument above is correct, we should also see climate change as an issue of human rights.

The argument for a rights-based approach to climate change is, however, far from complete. Some, for example, would argue that this right cannot be held by members of future generations (the extreme challenge). Others would argue that this right should be subject to a positive pure time discount rate and therefore the development rights of current generations may override the discounted rights of current and future people not to suffer from dangerous climate change (the moderate challenge). Space precludes examination of the extreme challenge here.[4] In the remainder of this chapter I focus on the moderate challenge. As we shall see, many reasons have been given (especially since the publication of the Stern Review) as to why it is appropriate to discount people's interests. In the light of this, I examine the case for pure time discounting. My claim is that the right that I have defended above (the right not to be exposed to dangerous climate change) should not be subject to a positive pure time discount rate. This allows the possibility that a discount rate might be applied to other values: it simply denies that rights should be discounted.

Rights and discounting

Let us turn now then to the claim that the intertemporal character of climate change is morally significant because the rights (or interests – for those who reject the claim

---

[4] For scepticism about the rights of future people see Beckerman and Pasek, (2001: 11–28). For excellent discussion of the rights of future people see Feinberg (1980: 180–3); Elliott (1989); Meyer (2003). See also Caney (2009b).

that future people have rights) of current and future generations should be subject to a positive "discount rate." That is to say the rights (or interests) of people should be ascribed less value the further they are into the future. On this view, the rights of future generations should be afforded less protection than the rights of contemporaries and, moreover, the rights of contemporaries should also be discounted throughout their life. By contrast, the position that I am defending affirms what is termed a zero discount rate to the protection of rights. On my account the rights of a person in the twenty-first century have the same moral standing as the rights of a person in the twenty-third century. This is an important conclusion because if we think that future people have rights of a lesser significance than the rights of contemporaries, then the rights of future people not to suffer from the ill-effects of climate change could more easily be overridden by the interests of the currently alive in activities which involve high levels of fossil fuel consumption. And indeed some object to large-scale policies of mitigation on precisely these grounds. So, whether we adopt a positive or a zero discount rate has great practical importance.[5]

Before considering challenges to my affirmation of a zero discount rate for rights, it is essential to make three preliminary points about discounting in general. First, we should distinguish between different accounts of what it is that should have a positive or zero (or negative) discount rate. What, we might ask, is the subject of the discount rate?[6] Two possibilities are "moral worth" and "resources." If we apply a positive discount rate to the "moral worth" we attribute to people then we hold that we should attribute less moral worth to people the later in time that they live. We exhibit what is termed "pure time preference." By the same token, if we apply a positive discount rate to the "resources" we allocate to people then we hold that we should spend less money on people the later in time that they live. Now the account that I am defending above is fundamentally committed to a zero discount rate for persons' moral worth (cf. Ramsey, 1928: 543; Pigou, 1946: 25–6; Sidgwick, 1981 [1907]: 414).[7] My aim is to defend the view that when determining persons' worth (and in particular when determining the importance of protecting people's rights) there should be no pure time preference. The two different kinds of discounting are often combined in the following very simple formula:

$$\text{Social discount rate} = \text{pure time preference } (\delta) + \eta \times (\text{rate of increase in consumption per capita}).[8]$$

---

[5] The importance of the discount rate is attested to by the response to the Stern Review's use of a low discount rate (2007: 35–7). For the responses which single out this issue as a key element in Stern's Review see Mendelsohn (2006–2007, especially pp. 42–3); Dasgupta, (2007); Nordhaus (2007); and Weitzman (2007).

[6] See Broome (1992: 52). Broome distinguishes between discounting for "well-being" and discounting for "economic commodities" (1992: 52).

[7] For a brilliant modern discussion of discounting see Parfit (1986: 480–6).

[8] This is the formula conventionally employed in the literature. See, for example, Stern (2007: 52). I thank Dominic Roser for helpful discussion.

My concern is with δ. My claim is that earlier rights should be accorded the same moral status as later rights: δ should be zero.

Second, we should note that those who affirm a positive value for δ vary in the kinds of discount rate they affirm. Some suggest a fixed social discount rate. They may propose, for example, that we should apply a pure time discount rate of 3% so those born now have a value of 1 and those born in 10 years have a value of $1/(1.03)^{10}$. One obvious upshot of a fixed pure time discount rate is that fairly soon the moral standing of members of future generations becomes very low. Partly in light of this, others suggest varying pure time discount rates which apply a higher discount rate to near future generations and lower rates to distant future generations. For example, some have advocated hyperbolic discount rates (on which see Ainslie (2001) and Groom *et al.* (2005: 471–3)). In a similar vein, Martin Weitzman has suggested a system of "sliding scale" discount rates, which divides the future into five separate periodizations (1–5 years, 6–25 years, 26–75 years, 76–300 years, and more than 300 years) and then allocates a different discount rate to each periodization (respectively, 4%, 3%, 2%, 1%, and 0%) (Weitzman 2001: 270). These variable rates have less dramatic effects on the moral weight of future generations so although they ascribe them less weight than current people they do not undercut the moral weight of future people so drastically. To take one important example, in their book *Warming the World*, William Nordhaus and Joseph Boyer (2000: 16) hold that the (moral) discount rate (i.e. the extent of pure time preference) should drop from 3% p.a. in 1995 to 2.3% p.a. in 2100 to 1.8% p.a. in 2200.

Third, it is useful to distinguish between two perspectives when thinking about the justice of discounting. We can consider matters from the perspective of the rights-bearer (are persons receiving their entitlements?), and also from the perspective of the duty-bearer (what duties do persons owe others?). Now looked at from a rights-bearer perspective it seems hard to see why there should be a positive pure time discount rate. The interests being invoked to explain why a present person has a right to protection from X would also explain, ceteris paribus, why any future person has a right to the *same* level of protection from X. Nothing in the argument developed in Section II gives us any grounds for thinking that the rights of some are of a lesser worth than the rights of others. Put more extravagantly we might say that a person is a person. Given this, it seems likely that arguments for a positive value for δ can succeed only if they work from a duty-bearer viewpoint. That is to say, the most plausible defence of a positive pure time discount rate will reflect in some way the interests of the putative duty-bearer.

What arguments might one adduce for either a fixed or a varying positive social discount rate? And do they undermine the view defended above? Consider four arguments.

Argument 1: the "argument from revealed preferences"

Some leading economists start from the assumption that persons exhibit myopia in their conduct – they prefer a pleasure to occur earlier rather than later – and infer from this that a social discount rate should, by the same token, also exhibit a partiality toward the present. William Nordhaus and Joseph Boyer (2000), for example, appear to affirm this line of reasoning. They use people's investment and savings behavior to ascertain people's attitude to discounting and then conclude that the extent of pure time preference should be reflected in government policy towards climate change.[9] They, thus, adhere to the following three assumptions:

1. Analyses of how to respond to the possibility of climate change should employ a discount rate for pure time preference that corresponds to the preferences of the people [the revealed preference assumption].
2. People's views on discounting for pure time preference can be ascertained by observing their behavior in the market place [the methodological assumption].

If we employ the method described in (2) we find that

3. People's behavior exhibits impatience [the empirical claim].

The correct approach to take towards climate change should therefore employ a positive pure time discount rate.

Each of these assumptions is, however, problematic. Consider (2). For many economists the appropriate method for ascertaining people's preferences is to observe their market behavior and, in particular, their decisions to invest, save, and spend. There are, at least, four problems with this. First, market mechanisms cannot cope with prisoners' dilemmas. Individuals may want a certain outcome but the dominant strategy may be to act in ways that will not bring about that outcome. Second, there are also "assurance" problems. These occur when people have preferences for a certain outcome but will not act to bring about that outcome unless they can be assured that others (whose action is also needed for the outcome to come about) will also contribute to the provision of the desired outcome. The problem is that market mechanisms cannot detect these preferences because they reflect individual preferences expressed separately. Political action can, however, overcome this coordination problem.[10] A third problem with (2) is that, as Mark Sagoff (1988)

---

[9] Nordhaus and Boyer defend their original discount rate of 3% on the grounds that it is "consistent with historical savings data and interest rates" (2000: 15). See also Nordhaus's rather confusing discussion of pure time preference in (1997). He begins by saying that the rate of time preference should be informed by people's actual behavior but then writes, quite correctly, that "we should be careful not to commit the naturalist fallacy of ethically equating what is with what should be" (1997: 317).

[10] For discussion of both of these two points see Sen's influential discussion of the "isolation" paradox and "assurance" problems in Sen (1961, especially p. 487ff, 1967, 1982: 328).

has long argued, people behave differently as consumers to the way they do as citizens. When people occupy different roles they often deploy, and should deploy, different kinds of reasoning. What persons choose in the market place sometimes differs from what they would choose in the ballot box. Market mechanisms are, therefore, incomplete as devices for fully reflecting people's preferences (Sagoff, 1988: 50–73; see also Sen, 1982: 328). Fourth, empirical analyses of people's attitudes towards time preference have shown that people do not have a single uniform approach to time. Rather they use different discount rates for different phenomena (depending, for example, on factors such as whether the good in question is a benefit or a loss, and depending on the magnitude of the benefit/loss) (Loewenstein and Thaler, 1989; Loewenstein and Prelec, 2000). Given this we have no reason to suppose that any pure time discount rate that people adopt in market exchanges would be the same pure discount rate that those same people would apply to climate change. Assumption (2) is thus highly problematic.[11]

Now in response to these objections some might argue that we could use other means for determining people's attitude to the future – for example, opinion polls, focus groups, referenda – and that these show that people exhibit myopia (Cropper *et al.*, 1994). Let us therefore consider (3). This claim may seem hard to contest but it is not, I submit, as straightforward as might first appear. As Shane Frederick (2003) has argued, whether people exhibit pure time preference or not depends on the method being employed. He identified six different elicitation procedures (that is methods for ascertaining how much people think we should discount for time) and then conducted a survey of 401 people. His research found that these six different procedures all issue in different results – some of them markedly different (Frederick, 2003). For example, when people were asked to compare a death from pollutants 100 years from now compared to a death from pollutants next year, 64% replied that they were "equally bad" (Frederick, 2003: 43). A similar result was obtained when people were asked whether they would prefer a policy that saved 300 lives in the current generation, 0 lives in the next generation, and 0 lives in the next generation after that to a policy that saved 100 lives in this generation, 100 lives in the next generation, and 100 lives in the generation after that. Frederick reports that 80% preferred the second policy (2003: 46). By doing so they chose a view that does not discriminate against future generations and they rejected a view that is characterized by pure time preference. The assumption that people's views are strongly in favor of positive pure time discounting is therefore not as straightforward as is often assumed. People exhibit different views depending on the kind of intertemporal issue in front of them, and how the options are described. People do not have a single undifferentiated approach to pure time preference. In addition to all the

---

[11] For similar points see Beckerman and Hepburn (2007: 203–204).

above, we should be circumspect in reaching conclusions based on the use of opinion polls and surveys to identify people's commitment to "pure time preference." The answers given to the questions posed in such surveys will almost inevitably reflect other non-time-related factors – such as uncertainty ("how can we be sure the government will save 300 lives later?") and optimism ("maybe a cure will be found in which case we should prioritize those who are ill now"). They are, thus, fraught with problems as a means of identifying people's commitment to pure time preference (Frederick, 2003: 49–50).

Let us turn now to assumption (1). Perhaps the most serious flaw in the argument under consideration concerns its assumption that the rate of discount for pure time preference should be determined by (current) people's preferences. Why should we accept this? Where there are problems which persist over a very long time span this approach has the effect that the preferences of *some* determine the prospects of *others*. One might accept time discounting in cases where A's preference for discounting affects only A, but this is clearly not the case with climate change and, as Thomas Schelling has long pointed out, one cannot move from the case of discounting within a person's life to determine whether discounting between generations is appropriate. That I might want a unit of pleasure to occur earlier in my life rather than later in my life does not establish that it is permissible that I enjoy a unit of pleasure in my life rather than that some future person enjoy that same quantity (Schelling, 1995: 396). To claim the contrary would just be a non sequitur. My preferences towards either my pleasure later in my life or towards the interests of future generations do not logically imply anything about the extent of my own obligations to future people.

Argument 2: the "argument from welfarism"

Given the limitations of the first argument's invocation of people's preferences, let us now consider a second argument which also makes reference to people's revealed preferences. One problem with the last argument is that it gives us no reason to follow the revealed preferences of those who are currently alive. The second argument seeks to remedy this defect. It argues that the state should further people's welfare and this is why it ought to satisfy their preferences. Proponents of this argument then argue that people have a preference for pleasures to take place sooner rather than later: they have a preference for "impatience." Therefore the state should adopt a positive discount rate. Kerry Turner, David Pearce, and Ian Bateman (1994) have, for example, reasoned in this way. They defend discounting on the basis that "people prefer to have benefits now rather than later" and that "the very rationale for CBA [cost–benefit analysis – SC] is that preferences count" (1994: 97; also Pearce *et al.*, 1989: 132–3, 2003: 122).

This argument is seriously flawed. It claims to follow from a commitment to welfarism but it fails to understand the nature of a welfarist political morality. It is problematic in four ways. First, welfarism holds that all people's preferences should count and should do so equally. These are its fundamental tenets. The aim of a (maximizing) welfarist is to satisfy as many preferences as possible. As such the preferences of those who will be born matter and there is nothing within the idea of welfarism to justify, or even permit, disregarding or marginalizing the preferences of some just because others think that they should count equally. Second, the egalitarian principle that each person's preferences should count equally is part of what gives welfarism its appeal. From Jeremy Bentham to Peter Singer, welfarists have stressed the egalitarian character of welfarism. As such a welfarist approach should bracket out preferences about whose preferences should count and by how much. Third, this argument is guilty of a non sequitur. It is, of course, correct that welfarists think that preferences about what goods or services to consume and which activities to engage in should be satisfied. However, it does not follow from this that the fundamental tenets of welfarism (its commitment to treating all equally and to maximization) should themselves be decided by preference satisfaction. To think this is to commit a category mistake. *Argument 2* is thus inadequate, even if we assume that all issues in political morality should be determined wholly from the perspective of cost–benefit analysis. However, we should also note here that the argument developed earlier defends a human-rights approach and denies that cost–benefit analysis is the whole of political morality. *Argument 2* thus does not bear on the argument defended. To say this is not to say that cost–benefit analysis should be rejected. It is the more modest point that there is more to political philosophy than cost–benefit analysis, that there are rights (including the right not to suffer from dangerous climate change), and hence that the second welfarist-inspired argument has no purchase on the discount rate affirmed here.

Argument 3: the "argument from demandingness"

The first two arguments start from a concern with satisfying people's preferences. As we have seen, both versions have failed. We turn now to a third argument. This argument too can be seen as being motivated by a concern for honoring people's preferences. Some contend that a zero pure time discount rate is unduly demanding on people. It can require people to make unreasonably large sacrifices in cases where doing so would benefit not just the next generation but all the ones after that. This troubling conclusion can be avoided if we posit a positive discount rate (either fixed or varying). A zero discount rate, however, leaves us committed to imposing heavy sacrifices on any generation to promote the wellbeing of its successors (Lomborg,

2001: 314; Pearce *et al.*, 2003: 124–5; Posner, 2004: 152–3).[12] As such, it denies current generations the space to pursue their own goals and to satisfy their own preferences.

Prior to evaluating this argument, note that this defense of a positive discount rate has the feature I suggested an argument for a positive discount rate should adopt – namely it reflects a duty-bearer perspective. Its complaint is that a zero discount rate is too onerous on duty-bearers.

Five points can be made in reply. First, we should distinguish between different kinds of behavior. Many distinguish between negative and positive duties, where the former refer to duties to abstain from certain actions (e.g. a duty not to murder) and the latter refer to duties to perform certain actions (e.g. a duty to aid). Now this distinction is relevant here because many argue that discounting is most plausible for positive duties and is implausible for negative duties. Put less abstractly, the suggestion is that the duty not to kill or the duty not to expose people to dangerous risks should not be subject to discounting, but positive duties are perhaps more vulnerable to discounting.[13]

A second even more telling point can also be made. Whether a principle of intergenerational justice is unduly demanding is a function not simply of whether there is no discounting but also of the content of the principle. Those pressing the objection assume that it is the role of the state to maximize preference satisfaction, and their argument then is that if maximal preference satisfaction is the goal, this will demand the imposition of heavy costs on early generations when doing so would satisfy the preferences of all or many following generations. The view defended above, however, is immune to this objection because it is not committed to the maximization of preference satisfaction. It applies a zero pure time discount rate to one specific value (the realization of people's basic rights) and this scope-restricted view is, therefore, not vulnerable to the argument from demandingness. To hold that the basic rights of all persons should be treated on an equal basis would not require the highly demanding sacrifices that a maximizing view (or indeed other views) might.[14]

Those who propound the argument from demandingness overlook the possibility of a scope-restricted view (that is, the view that a zero discount rate should apply to some but not all values). David Pearce and his co-authors, for instance, maintain that zero discounting is problematic because "[z]ero discounting means that we *care as much for someone* not just one hundred years from now as we do for someone now, but also

---

[12] Kenneth Arrow also makes a similar point (1999: 14–16). He invokes Samuel Scheffler's (1982) "agent-centred prerogative" to claim that each generation should favor itself and then treat all the others the same (1999: 16).

[13] For a related point see Parfit (1986: 486). See also Broome (1992: 107–108); Davidson (2006: especially pp. 59 & 66); de-Shalit (1995: 13–14 and 63–4).

[14] See also Broome (1992: 106). As he notes, this point has also been made by Parfit (1986: 484–5), and by Rawls (1999: 262), who argues that the problem is with a maximizing view and to add discounting is just an "ad hoc" solution.

someone one thousand years from now, or even one million years from now" (Pearce *et al.*, 2003: 124; emphasis added). But as I have argued above, zero discounting in itself does not, of necessity, entail this. The version of zero discounting defended in this chapter is quite consistent with people *caring more* for their contemporaries. It just insists that when it comes to fundamental rights, we should treat people on a par, independently of which generation they are born into, and then above that, they can devote more (indeed far more) care to those close to them. This approach disaggregates moral principles and maintains that some values (such as rights) behave differently to other values (such as loyalty or care). Our sentiments of loyalty or care may diminish over time. Our commitment to fundamental rights may not.

Three additional points also merit attention. We have seen that a basic rights view is not vulnerable to the concern that a zero discount rate is unduly demanding. Let us now consider the structure of argument from demandingness in more detail. Put in its barest form the charge is that *given*, (a) a commitment to maximum preference satisfaction, if we affirm, (b) a zero discount rate, then, (c) we would be committed to an intolerably demanding morality. The suggestion then is that we should abandon (b). But two points could be made against this. First, we need an argument as to why dropping (b) rather than (a) is the appropriate response. Why does this not call into question the commitment to maximal preference satisfaction?[15] Second, the case for abandoning (a) rather than (b) is strengthened further when we put the debates about utilitarianism, demandingness, and discount rates into context. Utilitarianism is often criticized for being unduly demanding – not just in the intergenerational context but also in intra-generational contexts. Discounting future generations may prevent utilitarianism from being unduly demanding in intergenerational contexts, but it does nothing to meet the charge that maximizing consequentialism asks too much in intra-generational contexts. Discounting over time thus remains a partial solution to a more general problem with maximal preference satisfaction.

One final observation is in order. In his version of the argument from demandingness, Kenneth Arrow appeals to Samuel Scheffler's concept of an "agent-centred prerogative" to defend discounting (Arrow, 1999: 16). Scheffler's claim is that each person is morally entitled to further their own interests and goals. They have what he terms an "agent-centred prerogative" to promote their own personal projects (Scheffler, 1982). But Scheffler's (very plausible) argument is that a commitment to an agent-centered prerogative claim entails that we should reject a maximizing consequentialism. His claim is *not* that the utility of other persons should be discounted and that we should retain a commitment to maximization. The whole point of his argument is that persons do not have a duty to maximize well-being.

---

[15] Again see the illuminating comments on this by Parfit (1986: 484–5); Broome (1992: 106); Rawls (1999: 262).

Hence the title of his book: *The Rejection of Consequentialism* (Scheffler, 1982). To invoke agent-centred prerogatives while also affirming a maximizing conse-quentialism thus misunderstands Scheffler's argument and the conclusions that he plausibly derives from it. This, however, is precisely what Arrow does, for after invoking Scheffler's agent-centered prerogative he holds that society should "*maximize* a weighted sum of its own utility and the sum of utilities of all future generations, with less weight on the latter" (Arrow, 1999: 16; emphasis added).

Argument 4: the "argument from economic growth"

Any analysis of the case for discounting would be incomplete without considering a fourth and final argument for discounting that is often adduced. Like the previous argument, it thinks that discounting is unfair to current generations, but it takes a different tack. It defends discounting on the grounds that there will be economic growth and future generations will be wealthier than current generations. It then argues that the more wealth persons have, the less marginal utility results from each extra increment of wealth (i.e. the law of diminishing marginal utility). Hence it makes sense to employ a positive discount rate: wealth produces more utility if it is spent on earlier rather than later generations. It would therefore be wrong not to discount.

This argument is, however, also unsuccessful as a critique of the view defended here. It is flawed for at least two reasons. The first problem is apparent once we recall the distinction between discounting persons' "moral worth," on the one hand, and discounting "resources," on the other. I have defended a zero discounting approach to persons' moral worth (and in particular to their rights), but the third argument simply does not speak to this. Its focus is on how many "resources" should be allocated to present and future people, and it makes a case for spending more sooner rather than later. Nothing that it says, however, can show that the rights (or indeed interests) of future people are of a lesser worth and should be discounted. All it says is that spending resources on earlier generations will result in higher preference satisfaction (because each unit of wealth will boost preference satisfaction when spent on poorer rather than richer people), and it cannot establish that the interests of future people are less valuable than those of present people. To put the same point in a different way: one can accept this argument and still hold that the rights (and interests) of future people not to suffer dangerous climate change should be treated as having the same weight as the rights (and interests) of present people.

Second, suppose that we set the first point aside. Even then the argument is flawed. For the argument assumes that future people will be wealthier than current people. However, as Schelling notes it is important to disaggregate future gener-ations and not treat them as a unitary group (1995: 398–400). It might be true that future members of some countries (e.g. the United States or Britain) will be wealthier than the current members of those same countries. This, however, does not establish that an analysis of global climate change should employ a positive social discount rate. Consider, for example, some of the likely victims of climate change – take Bangladeshis of the twenty-second century. It seems highly likely that they will be poorer than the current inhabitants of the United States. In this kind of case the argument from diminishing marginal utility clearly cannot vindicate a positive social discount rate. One cannot then defend a policy of discounting the interests of the future people affected by climate change on the grounds that they will be wealthier than those who will currently bear the burdens, for that it is simply not true of some future people. So the fourth argument does not even support a positive discount rate for spending resources. In fact, if future Bangladeshis are poorer than contemporary Americans, there should be a negative discount rate.[16]

So to conclude this section, we can see, first, that the argument adduced in support of a right not to suffer from climate change entails that that right is held equally by all. Its premises contain in them no room for a positive pure time discount rate. Furthermore (and this is the second point) we have seen that the four challenges that we have considered fail to undermine this view.

Conclusion

Some affirm a very minimal set of rights and would be sceptical of extending this set to include "environmental" rights of any kind. Others do not take such a hostile approach, but do ask why we should accept a right to a safe environment. The Stern Review, for example, insists quite rightly that rights "should be argued rather than merely asserted" (Stern, 2007: 47). I hope to have provided such an argument in this chapter. The kinds of considerations that we normally invoke to defend human rights, I maintain, entail that persons have a human right not to suffer from the ill effects of global climate change. Climate change undermines persons' human rights to a decent standard of health, to economic necessities, and to subsistence.[17] I have, moreover, argued that this right should not be discounted. Its moral importance does not diminish over time. In doing so, however, I have defended a scope-restricted view with respect to discounting. That is to say, I have defended a view which (1) holds that basic rights should not be discounted, but (2) allows for the possibility that that other values might be subject to a positive pure time discount rate.

---

[16] For further discussion see Caney (2009c).

[17] Note that I am *not* claiming that this is the only reason why anthropogenic climate change is morally unacceptable. My arguments here are compatible with many other lines of critique.

More needs to be done – including, most notably, providing an analysis of who should bear the burdens of global climate change – before we have a complete theory of climate justice (Caney, 2005, 2009a, 2009b). In the meantime, however, I hope that I have provided the beginnings of an argument for the injustice of global climate change.

Acknowledgements

This chapter is a revised version of "Climate Change, Human Rights and Discounting," *Environmental Politics*, **17**(4) July 2008, pp. 536–55. The research for this paper was conducted as part of an Arts and Humanities Research Council research project on "Global Justice and the Environment" and the paper was written during my tenure of a Leverhulme Research Fellowship. The final version was revised while I held an ESRC Climate Change Leadership Fellowship. I thank the AHRC, ESRC, and Leverhulme Trust for their support. Earlier versions of this chapter were presented at Nottingham University (February 19, 2007); Amsterdam University (March 22–23, 2007); Liverpool University (April 19–20, 2007); University of Washington (May 3–4, 2007); Oxford Brookes University (May 30, 2007); University of Oslo (June 21–24, 2007); Manchester University (June 27–28, 2007); and Oxford University (September 21–22, 2007). I am grateful to those present for their questions, and thank, in particular, Michael Blake (who was my commentator at the University of Washington) and an anonymous referee for *Environmental Politics*. I would also like to thank Neil Adger, Paul Baer, Derek Bell, Richard Dagger, Steve Gardiner, John Harris, Ellen Hey, Mathew Humphrey, Dale Jamieson, Andrew Light, Joan Martinez Alier, David Miller, Kieran Oberman, Karen O'Brien, Roland Pierik, Jonathan Quong, Matthew Rendall, Dominic Roser, Steve Schneider, Henry Shue, Adam Swift, David Stevens, and Wouter Werner.

References

Adger, W. N. 2004. The right to keep cold. *Environment and Planning A*, **36**(10), 1711–15.
Ainslie, G. 2001. *Breakdown of the Will*. Cambridge, UK: Cambridge University Press.
Arrow, K. 1999. Discounting, morality, and gaming. In P. R. Portney and J. P. Weyant, eds., *Discounting and Intergenerational Equity*. Washington, DC: Resources for the Future, pp. 13–21.
Azar, C. and Schneider, S. H. 2002. Are the economic costs of stabilizing the atmosphere prohibitive? *Ecological Economics*, **42**(1), 73–80.
Barker, T. *et al.* 2007. Summary for Policymakers. In B. Metz, O. R. Davidson, P. R. Bosch, R. Dave and L. A. Meyer, eds., *Climate Change 2007. Mitigation: Contribution of Working Group III to the Fourth Assessment Report of the Intergovernmental Panel on Climate Change*. Cambridge, UK: Cambridge University Press, pp. 1–23.

Beckerman, W. and Hepburn, C. 2007. Ethics of the discount rate in the Stern review on the economics of climate change. *World Economics*, **8**(1), 187–210.

Beckerman, W. and Pasek, J. 2001. *Justice, Posterity, and the Environment*. Oxford, UK: Oxford University Press.

Broome, J. 1992. *Counting the Cost of Global Warming*. Cambridge, UK: White Horse Press.

Caney, S. 2005. Cosmopolitan justice, responsibility and global climate change. *Leiden Journal of International Law*, **18**(4), 747–75.

Caney, S. 2009a. Global climate change and the duties of the advantaged. *Critical Review of International Social and Political Philosophy*, **12**(4), 693–717.

Caney, S. 2009b. Human rights, responsibilities and climate change. In C. Beitz and R. Goodin, eds., *Global Basic Rights*. Oxford, UK: Oxford University Press, pp. 227–47.

Caney, S. 2009c. Climate change and the future: time, wealth and risk. *Journal of Social Philosophy*, **40**(2), 163–86.

Caney, S. 2009d. Climate change, human rights and moral thresholds. In S. Humphreys, ed., *Human Rights and Climate Change*, Cambridge, UK: Cambridge University Press, pp. 69–90.

Confalonieri, U. *et al.* 2007. Human health. In M. L. Parry, O. F. Canziani, J. P. Palutikof, P. J. van der Linden and C. E. Hanson, eds., *Climate Change 2007: Impacts, Adaptation and Vulnerability. Contribution of Working Group II to the Fourth Assessment Report of the Intergovernmental Panel on Climate Change*. Cambridge, UK: Cambridge University Press, pp. 391–431.

Cropper, M. L., Aydede, S. K. and Portney, P. R. 1994. Preferences for life saving programs: how the public discounts time and age. *Journal of Risk and Uncertainty*, **8**(3), 243–65.

Dasgupta, P. 2007. Commentary: The Stern review's economics of climate change. *National Institute Economic Review*, **199**(1), 4–7.

Davidson, M. D. 2006. A social discount rate for climate damage to future generations based on regulatory law. *Climatic Change*, **76**(1/2), 55–72.

de-Shalit, A. 1995. *Why Posterity Matters: Environmental Policies and Future Generations*. London: Routledge.

Elliott, R. 1989. The Rights of Future People. *Journal of Applied Philosophy*, **6**(2), 159–70.

Feinberg, J. 1980. The rights of animals and unborn generations. In *Rights, Justice, and the Bounds of Liberty: Essays in Social Philosophy*. Princeton, NJ: Princeton University Press, 159–84.

Frederick, S. 2003. Measuring intergenerational time preference: are future lives valued less? *Journal of Risk and Uncertainty*, **26**(1), 39–53.

Groom, B., Hepburn, C., Koundouri, P. and Pearce, D. 2005. Declining discount rates: the long and the short of it. *Environmental and Resource Economics*, **32**(4), 445–93.

Hare, B. 2006. Relationship between increases in global mean temperature and impacts on ecosystems, food production, water and socio-economic systems. In H. J. Schellnhuber, W. Cramer, N. Nakicenovic, T. Wigley and G. Yohe, eds., *Avoiding Dangerous Climate Change*. Cambridge, UK: Cambridge University Press, pp. 177–87.

Hayward, T. 2005. *Constitutional Environmental Rights*. Oxford, UK: Oxford University Press.

Kovats, R. S., Campbell-Lendrum, D. and Matthies, F. 2005. Climate change and human health: estimating avoidable deaths and disease. *Risk Analysis*, **25**(6), 1409–18.

Loewenstein, G. and Prelec, D. 2000. Anomalies in intertemporal choice: evidence and an interpretation. In D. Kahneman and A. Tvsersky, eds., *Choices, Values, and Frames*. Cambridge, UK: Cambridge University Press, pp. 578–96.

Loewenstein, G. and Thaler, R. 1989. Anomalies: intertemporal choice. *Journal of Economic Perspectives*, **3**(4), 181–93.

Lomborg, B. 2001. *The Skeptical Environmentalist: Measuring the Real State of the World*. Cambridge, UK: Cambridge University Press.

McMichael, A. J. *et al.*, eds. 2003. *Climate Change and Human Health: Risks and Responses*. Geneva: World Health Organization.

McMichael, A. J. *et al.* 2004. Chapter 20: Global Climate Change. In M. Ezzati, A. D. Lopez, A. Rodgers and C. J. L. Murray, eds., *Comparative Quantification of Health Risks: Global and Regional Burden of Disease Attribution to selected Major Risk Factors*. Geneva: World Health Organization, pp. 1543–649.

Mendelsohn, R. 2006–2007. A Critique of the Stern Report. *Regulation* **29**(4), 42–6.

Meyer, L. H. 2003. Past and future: the case for a threshold notion of harm. In L. H. Meyer, S. L. Paulson and T. W. Pogge, eds., *Rights, Culture, and the Law: Themes from the Legal and Political Philosophy of Joseph Raz*. Oxford, UK: Oxford University Press, pp. 143–59.

Nickel, J. 1993. The human right to a safe environment: philosophical perspectives on its scope and justification. *Yale Journal of International Law*, **18**(1), 281–95.

Nordhaus, W. D. 1997. Discounting in economics and climate change. *Climatic Change*, **37**(2), 315–28.

Nordhaus, W. D. 2007. A review of the Stern review on the economics of climate change. *Journal of Economic Literature*, **45**(3), 686–702.

Nordhaus, W. D. and Boyer, J. 2000. *Warming the World: Economic Models of Global Warming*. Cambridge, MA: The MIT Press.

Page, E. 2006. *Climate Change, Justice and Future Generations*. Cheltenham, UK: Edward Elgar.

Parfit, D. 1986. *Reasons and Persons*. Oxford, UK: Oxford University Press.

Patz, J. *et al.* 2000. The potential health impacts of climate variability and change for the United States: Executive Summary of the Report of the Health Sector of the US National Assessment. *Environmental Health Perspectives*, **108**(4), 367–76.

Pearce, D., Groom, B., Hepburn, C. and Koundouri, P. 2003. Valuing the future: recent advances in social discounting. *World Economics*, **4**(2), 121–41.

Pearce, D., Markandya, A. and Barbier, E. B. 1989. *Blueprint for a Green Economy*. London: Earthscan.

Pigou, A. C. 1946. *The Economics of Welfare*. 4th edn. London: Macmillan.

Posner, R. 2004. *Catastrophe: Risk and Response*. Oxford, UK: Oxford University Press.

Ramsey, F. 1928. A mathematical theory of saving. *Economic Journal*, **38**(152), 543–59.

Raupach, M. R. *et al.* 2007. Global and regional drivers of accelerating CO_2 emissions. *Proceedings of the National Academy of Sciences of the United States of America*, **104**(24), 10 288–93.

Rawls, J. 1999. *A Theory of Justice*, revised edn. Oxford, UK: Oxford University Press.

Raz, J. 1986. *The Morality of Freedom*. Oxford, UK: Clarendon Press.

Sagoff, M. 1988. *The Economy of the Earth*. Cambridge, UK: Cambridge University Press.

Scheffler, S. 1982. *The Rejection of Consequentialism: A Philosophical Investigation of the Considerations Underlying Rival Moral Conceptions*. Oxford, UK: Oxford University Press.

Schelling, T. 1995. Intergenerational discounting. *Energy Policy*, **23**(4/5), 395–401.

Sen, A. K., 1961. On optimizing the rate of saving. *Economic Journal*, **71**.

Sen, A. K., 1967. Isolation, assurance and the social rate of discount. *The Quarterly Journal of Economics*, **81**(1), 112–24.

Sen, A. K. 1982. Approaches to the choice of discount rates for social benefit–cost analysis. In Robert C. Lind *et al.*, eds., *Discounting for Time and Risk in Energy Policy*. Washington, DC: Resources for the Future, pp. 325–53.

Sidgwick, H. 1981 [1907]. *The Methods of Ethics*, 7th edn. Indianapolis, IN: Hackett Publishing Company.

Stern, N. 2007. *The Economics of Climate Change: The Stern Review*. Cambridge, UK: Cambridge University Press.

Solomon S. *et al.* 2007. Technical summary. In S. Solomon *et al.*, eds., *Climate Change 2007: The Physical Science Basis. Contribution of Working Group I to the Fourth Assessment Report of the Intergovernmental Panel on Climate Change*. Cambridge, UK: Cambridge University Press, pp. 20–91.

Turner, R. K., Pearce, D. and Bateman, I. 1994. *Environmental Economics: An Elementary Introduction*. New York, NY: Harvester Wheatsheaf.

Weitzman, M. L. 2001. Gamma discounting. *American Economic Review*, **91**(1), 260–71.

Weitzman, M. L. 2007. A review of the Stern review on the economics of climate change. *Journal of Economic Literature*, **45**(3), 703–24.

8

Climate change as a global test for contemporary political institutions and theories

STEPHEN M. GARDINER

If political leaders have one duty above all others, it is to protect the security of their people ... And yet our long-term security is threatened by a problem at least as dangerous as chemical, nuclear or biological weapons, or indeed international terrorism: human-induced climate change.

Houghton (2003)

Why should political philosophy be concerned about global environmental change, in general, and climate change, in particular? Why aren't these just normal political problems, perplexing in their scale, perhaps, but not fundamentally different to most other problems in domestic and international affairs? Why isn't the political problem, insofar as there is one, simply that certain actors have behaved badly, for the usual political reasons? This chapter is an attempt to offer one central answer to such questions.[1] It does so by advancing a minimal global test for social and political institutions and theories, and then suggesting that conventional versions of both may fail in the case of climate change. If this argument is correct, then climate change presents a major challenge to global systems.[2] This implies that the current, almost exclusive, focus on scientific and economic questions is a dangerous mistake.

The structure of the chapter is as follows. The first section, "The global test," introduces the global test[3] and provides some general reasons for believing that it may

---

[1] These are questions that I am often asked, but I am grateful to Stephen Macedo for pressing them most forcefully.

[2] For convenience, I will sometimes use the phrase "a global system" to refer to a set of global social and political institutions (including states and other subnational institutions), and the philosophies that support them.

[3] The term "global test" I take from Senator John Kerry, who invoked it in the 2004 presidential election in the United States, in a criticism of President George W. Bush. Senator Kerry had in mind the need to consult with other countries about security matters, and to convince them of real threats, as a way of maintaining American influence in world affairs. The President subsequently ridiculed the Senator's claim, arguing that the basic security of the United States should not be made conditional on the opinions of other nations. The test I have in mind here is different. It does, however, tend to suppose that the security of any state is dependent to some extent on the security of the global system of which it is part, and that this implies that state sovereignty *may* not be absolute.

Climate Change, Ethics and Human Security, eds. Karen O'Brien, Asunción Lera St.Clair and Berit Kristoffersen. Published by Cambridge University Press. © Cambridge University Press 2010.

apply in the case of climate change. The next section, "Scenarios," tries to say more precisely what is so worrying about climate change, by identifying two challenges to institutions and theories – the Hard Landing and Crash Landing scenarios – that it may bring on. The section following this advances the conjecture that existing global systems are poorly placed to handle such scenarios, and argues that humanity's initial response to the climate crisis appears to confirm this conjecture. The "Theoretical vices" section some basic difficulties for evaluating political theories in this setting, and tries to address them by pointing out some vices such theories may have. Finally, the last section illustrates the relevance of these characteristic vices through a brief discussion of utilitarianism and cost–benefit analysis.

It is perhaps worth emphasizing at the outset that, although the argument of this chapter is primarily negative, the motivation is not to disparage contemporary institutions and theories, many of which have been very useful for other purposes, but to advance them. The case of climate change helps us to see ways in which our systems (of thought and action) may need to be reoriented. As global ethics emerges as a major concern in both political philosophy and the world at large, this is an important task.

The global test

In July 2003, Sir John Houghton, former co-chair of scientific assessment for the United Nations' Intergovernmental Panel on Climate Change (IPCC), published an open letter to US President, George W. Bush, and British Prime Minister, Tony Blair, in the British press. Frustrated with the lack of action on climate change, Houghton accused the two leaders of neglecting their fundamental political duty towards their citizens in "an abdication of leadership of epic proportions" (Houghton, 2003). Houghton's charge is remarkable for two reasons. First, this is a very serious accusation for one senior public figure to make against two others, especially when the accusing is done in such a public way. This is important in its own right, since it suggests that Houghton regards the stakes as being very high. Second, Houghton's language clearly suggests the belief that from the political point of view there is something special about the threat posed by climate change. He is not, it seems, regarding climate change as a "normal" kind of political problem. Instead, in couching his complaint in terms of the fundamental duty of political leaders, he implies that there is something deep and basic about it.

Houghton's charge has intuitive appeal. Still, one might doubt whether he himself pushes it far enough. On the surface at least, Houghton seeks only to put climate change on a similar footing to issues such as international terrorism and the spread of weapons of mass destruction. These are serious problems to be sure. Yet many people, including some mainstream politicians, argue that climate change is pre-eminent among them. Moreover, the scope of Houghton's charge also seems too narrow.

Focusing on President Bush and Prime Minister Blair makes the complaint appear personal and isolated from any wider political context. But even if these two individuals should take some (central) responsibility for past international neglect of climate change, surely there are other contributors to the problem. In particular, not only had the issue been around for much longer than their administrations, but many political leaders seemed to have agreed with Al Gore's statement, from early in his term as Vice President of the United States, that "the minimum that is scientifically necessary [to combat global warming] far exceeds the maximum that is politically feasible" (McKibben, 2001: 38). In essence, the complaint is that the inaction of our leaders merely reflects wider political realities. If this complaint is justified, then the concern that Houghton identifies runs deeper than he himself implies. In short, the worry arises that the charge of fundamental failure can be leveled not just against particular leaders or administrations, but also more generally, against current social and political institutions, and the mainstream moral and political theories that support them.

Such thoughts motivate the global test. Suppose that human life on this planet was subject to some serious threat. Moreover, suppose that this threat was caused by human activities, but also preventable by changes in those activities. Add to this that the existing social and political systems had allowed the threat to emerge and then shown themselves to be incapable of adequately responding to it. Would this failure to act license a criticism of the existing social and political systems? If so, how serious a criticism would this be?

Suppose that the fact of global failure would indeed count as a criticism of existing systems and that such a criticism is potentially fatal. Moreover, assume that the charge of global failure can be applied not only to social and political institutions, but also to the philosophies that stand behind them. Under these assumptions, we seem to have identified an important global test for social and political institutions and theories: if either does not respect the claim that failure to address a serious global threat is a criticism of it, and a potentially fatal one, *then it is inadequate and must be rejected.*[4]

On the face of it, this is an important claim. The global test functions as a condition of adequacy for institutions and theories; it sets a constraint on their acceptability.[5] At first, stated baldly and at this high level of abstraction, the test may

---

[4] My project has some structural similarity to Dryzek (1987). But it differs from Dryzek's in several respects, including its scope (his concern is exclusively with ecological problems; mine is wider), its targets (his are social choice mechanisms; mine include political theories), its critical diagnosis (he blames "instrumental rationality"; I take no position), and its framework for solutions (he claims that we must move to more discursive and decentralized decision-making institutional bodies; I emphasize theoretical change and make no institutional claims here).

[5] Note that although the global test indicates one serious constraint on global political philosophy, it need not be the only such constraint, nor the dominant one. Indeed, this seems unlikely. After all, the test itself is narrowly conceived (e.g., because it deals only with self-generated threats), and there are other important areas of social and political concern (such as individual rights, distributive justice, intergenerational justice, the preservation of communities, our relationship to nature, and so on).

appear so obvious and unexceptional as to be barely worth mentioning.[6] But, as we shall now see, this appearance is deceptive. First, the test is highly relevant to current concerns, since a strong prima facie case can be made that climate change fulfils the basic conditions suggested in the schematic example, and so constitutes a case of global failure. Second, so far mainstream discussions of the climate problem – in politics, academia, and society at large – have largely ignored the test. Instead, the discourse is dominated by scientific, economic, and (short-term) geopolitical concerns, and comparatively little has been said about the adequacy of existing social and political systems. In short, the concern highlighted by the test is conspicuous by its absence from contemporary debates. Third, this fact should give us pause. As we shall see, one way of failing the test is to be oblivious, complacent, or even evasive about its concerns. In our current setting, this is a real worry. Earlier we saw that Sir John Houghton accused President Bush and Prime Minister Blair of "an abdication of leadership of epic proportions." Is it possible that our institutions and theories are vulnerable to the same charge?

Scenarios

Climate change is a difficult moral problem. There are many reasons for this. Elsewhere I have highlighted three of them: that it is genuinely global; that it is intergenerational; and that we are poorly equipped to deal with it theoretically (Gardiner, 2004a, 2009). In addition, I have argued that the convergence of these factors both puts us in an especially bad moral situation that I call a "perfect moral storm," and that this situation makes us vulnerable to moral corruption, including the corruption of our moral and political theories (Gardiner 2006a, in press). Let me begin by briefly reviewing this account.

The global challenge (or "storm") is familiar. Both the sources and the effects of anthropogenic emissions are spread throughout the world, across local, national, and regional boundaries. According to many writers, this creates a tragedy of the commons situation, because the global system is not currently set up to govern this kind of commons. Worse, there are skewed vulnerabilities: those who are most vulnerable and least responsible will probably bear the brunt of climate change impacts, at least in the short to medium term. In contrast, the developed nations are, by and large, responsible for the bulk of emissions to this point, and appear much less vulnerable to the more immediate impacts than the less developed countries,

---

[6] The high initial level of abstraction is useful for our purposes, for two reasons. The first is that the abstract statement of the test leaves some latitude for competing traditions and political philosophies to offer different interpretations of its crucial terms. This is important because it reduces the risk that the basic formulation begs the question against some particular approach. The second reason is that, even when expressed in extremely abstract terms, the test retains some intuitive bite. There seem to be clear cases where almost everyone would agree that the global test is violated; and this suggests that it can be useful even when its precise details are left unexplored.

where most of the world's poor reside. This mismatch of vulnerability and responsibility is exacerbated by the fact that the developed countries are more powerful politically and more capable of impeding or bringing about a solution, but the less developed are poorly placed to call them to account.

The intergenerational challenge is less familiar. The impacts of climate change are subject to major time lags, implying that a large part of the problem is passed on to the future. One reason for this is that emissions of the main anthropogenic greenhouse gas, carbon dioxide, persist in the atmosphere for very long periods of time: the typical carbon dioxide molecule remains in the atmosphere for several hundred years, but 10–15% remains for 10 000 years, and 7% for 100 000 years. Consequently, the full cost of any given generation's emissions will not be realized during that generations' lifetime. This suggests that each generation faces the temptation of intergenerational buck-passing: it can benefit from passing on the costs and/or harms of its behavior to future people, even when this is morally unjustified. Moreover, if the behavior of a given generation is primarily driven by its concerns about what happens during its own lifetime, then such overconsumption is likely (see Gardiner, 2004b).

The third challenge is theoretical. We do not yet have a good understanding of many of the ethical issues at stake in global warming policy. For example, we lack compelling approaches to issues such as scientific uncertainty, international justice, intergenerational justice, and the appropriate form of human relationships to animals and the rest of nature. This causes special difficulties given the presence of the other storms. In particular, given the intergenerational storm and the problem of skewed vulnerabilities, each generation of the affluent is susceptible to arguments for inaction (or inappropriate action) that shroud themselves in moral language but are actually weak and self-deceptive. In other words, each generation of the affluent is vulnerable to moral corruption: if they give undue priority to what happens within their own lifetimes, they will welcome ways to justify overconsumption and so give less scrutiny than they ought to arguments that license it. Such corruption is easily facilitated by the theoretical storm and obscured by other features of the global storm.

My concerns about the perfect moral storm still stand. Here, however, my focus is more limited and concerns the theoretical issue alone. Climate change involves the intersection of a number of characteristics that conventional approaches to public policy are not well equipped to handle, such as uncertainty, the very long term, and the creation of different preferences and persons. Moreover, it integrates them in a mutually reinforcing way. Given this, it is not surprising that the problem exposes some weaknesses of current orthodoxy. This general theoretical challenge is serious. Still, the idea of the global test suggests something more specific. After all, other policy problems may involve similar convergences and reinforcement. For example, if a society is designing an appropriate set of family-leave policies for parents of babies and young children, it will face choices that have uncertain effects, and involve long-term considerations and reproduction issues. Indeed, perhaps this is true for almost all large-scale projects with long time horizons. Still, such projects do not (normally) pose a challenge to political practices of the form I want to discuss here. Instead, Houghton's remark suggests that there is a more specific reason that climate change is theoretically important. There is something special about climate change that raises fundamental questions about conventional social and political practices – something to do with human security.

This is why it makes sense to invoke the global test in the case of climate change, but not in many others with some of the same characteristics. Given this, it is natural to ask: why is climate change special? In order to start answering this question, we must first take a step backwards and attempt to clarify what we are discussing. One difficulty in talking about environmental issues, in general, and climate issues, in particular, *as such* is that both "environment" and "climate" are large "catch all" terms. Hence, in order to discuss the nature of the difficulty posed by climate change, it will be helpful to begin by distinguishing specific aspects of environmental and climate change.

Suppose we begin, somewhat roughly and artificially, with the idea that climate change (and environmental change more generally) is usually caused by inputs to physical and ecological systems which bring about alterations in those systems, and then cause impacts on humans, animals, plants, and places that they value. In the area of alterations of basic systems, change has a number of important dimensions. One dimension is the *magnitude of the increments* of change, which may be small, medium, large, or massive. A second dimension is *timing*. This includes matters such as the *speed* (e.g. slow, fast) and temporal *profile* of the alterations (e.g. even, bounded, bumpy, abrupt). At one extreme, change may be slow and involve evenly distributed physical effects. But at another extreme it may also be fast and abrupt, as for example, if there are significant thresholds in the climate system, the breaching of which causes significant disruption to normal processes. A third dimension is *scope*. The salient level of a particular climate change may be local, national, regional, or global; and the physical effects of such a change may also be predominantly realized at one or other of these various levels. For example, a collapse of the thermohaline circulation in the North Atlantic might be best understood primarily as a regional climate change even if it has significant effects on global physical processes (e.g. precipitation in some parts of Africa and Asia).

Integrating these first three dimensions can help us to make some useful categorizations. For present purposes, let us isolate four especially salient types of physical change:

• Creeping change: Slow and even change in small increments that is local in scope.
• Methodical change: Moderately paced and bounded change in medium increments that is national in scope.

- Dramatic change: Moderately paced and bumpy change in large increments that is global in scope.
- Spectacular change: Fast and abrupt change in massive increments which is global in scope.

The fourth dimension of climate change worth noting is the *extent* of these impacts.[7] For one thing, their *magnitude* may range from very minor to significant, major, or extraordinary. For another, the *valence* of the impacts is important: the effects may be positive, negative, or mixed. It is important to specify that the main reason that we care about climate change is because of its potential impacts on humans and other forms of life. Although we may have some interest in the physical and ecological effects of climate change in their own right, we are predominantly concerned with their implications for human, animals, plants, and places of special value to them.[8] Because we are concerned with possible failures of the global test, negative effects – ranging from the merely inconvenient to the catastrophic – will be our focus here. However, it is also true that some systems may have difficulty in dealing with some kinds of effect that are, considered in isolation, very positive. After all, it is possible that even a change that is, all things considered, a very good thing may impose high transition costs on society at large, or on some particular groups. This may be especially likely if the change is widespread and fundamental.

The fifth dimension concerns the *character* of the impacts of climate change. Are they reversible or irreversible? Are there readily available substitutes for what is lost or is it non-substitutable? Are the costs of adapting to the new situations high or manageable? For convenience, I will lump these issues together under the heading "*malleability*." The idea here is that our concern is with how well we can accommodate the effects of climate change on human and nonhuman systems. For example, effects that can be easily and cheaply reversed or softened through the availability of substitutes exhibit high malleability; whereas effects for which reversal or substitution would be very expensive, or even impossible, exhibit low malleability.

The point of this classification exercise is to allow us to distinguish four especially salient change scenarios: (1) Soft Landing (creeping change with significant,

---

[7] The scope of impacts will also vary. For simplicity I assume here that this is approximately the same as the scope of physical effects. But this need not be true, given the complexity of global social and political systems (especially the economic system).

[8] For this reason, there would be some rationale for omitting the dimensions of the physical effects considered merely as such from the taxonomy, since many will say that their relevance depends exclusively on their implications for impacts. I have chosen to leave them in here for two reasons. First, much of the scientific work does still revolve around physical effects rather than impacts, and it is worth keeping note of the fact that any claim about the connection between these two needs to be established separately. Second, some people will be concerned about physical impacts for reasons other than, and in addition to, their concern for human (and even other forms of) life. For example, some will regard effects on particular places, or the transformative anthropogenic influence more generally, as something to be deplored. See, for example, McKibben (1989).

Table 8.1 *Salient change scenarios*

	Soft landing	Rough landing	Hard landing	Crash landing
Change	Creeping	Substantial	Dramatic	Spectacular
Size	Small	Medium	Large	Massive
Speed	Slow	Medium	Medium	Fast
Temporal Profile	Even	Bounded	Bumpy	Abrupt
Salient Scope	Even	Bounded	Bumpy	Abrupt
Impacts				
Valence	Negative	Negative	Negative	Negative
Salient Scope	Local	National	Global	Global
Magnitude	Significant	Major	Severe	Extraordinary
Malleability	High	Moderate	Poor	None

but highly malleable, negative impacts); (2) Rough Landing (substantial change with major, and moderately malleable, negative impacts); (3) Hard Landing (dramatic change with severe, and poorly malleable, negative impacts); and (4) Crash Landing (spectacular change with catastrophic negative impacts with no malleability). These scenarios are summarized in Table 8.1.

The conjecture

Identifying these scenarios enables us to discuss different possible threats that may be posed by climate change. This is useful for a number of reasons. One reason is that a failure to make such distinctions often obscures what is at stake in debates about climate policy. Notice, for example, that now that outright scepticism about climate change science is much less fashionable, those who oppose a substantial response to the threat often do so on the back of the assumption that the threat posed by climate change is of the Soft Landing sort, whereas those who are most concerned about climate change are often thinking primarily of Hard Landing or Crash Landing scenarios. Still, the main purpose here is merely to allow us to put forward the following conjecture for consideration: even if we suppose that conventional institutions and theories might do reasonably well with addressing Soft Landing scenarios, there is little reason for confidence as we move towards the Hard and Crash Landing scenarios.

The point of the conjecture is this. Remember that we were trying to understand why climate change might pose a special challenge to political systems and philosophies akin to the fundamental failure Sir John Houghton attributes to political leaders. I claimed that "climate change" (like "environmental change") is a large "catch all" term, and that this meant that we would need to make some distinctions. Having identified several different kinds of climate change (Creeping, Substantial,

Dramatic, and Spectacular Change) and a variety of different threat scenarios that might emerge from these (Soft, Rough, Hard, and Crash Landings), the conjecture then asserts that, although Soft Landing scenarios might pose no special problem for conventional institutions and theories, the Hard and Crash Landing scenarios do. Suppose then that our ethical concern is primarily with Hard Landing, Crash Landing, and any variety in between. The suggestion then emerges that if conventional political institutions and theories are poor at responding to such scenarios, then the global test implies that they are to be criticized for that. Moreover, if this problem is deep – for example, if it turns out that they *cannot* respond adequately – then they fail the test outright.

Why might one accept the conjecture? As it stands, it is quite general and applies regardless of the ideal strategy for dealing with the particular problem at hand. We might, however, refine the discussion by considering a variety of strategies for dealing with change. Suppose, for simplification, that we assume that there are two ways of responding to a potential change: those that involve addressing the cause; and those that involve addressing the effects. Consider, first, three basic strategies for dealing with the cause of a potential change. First, one might try to eliminate the cause, so that the effect does not arise (call this "Prevention"). Second, one might try to reduce the magnitude or scope of the cause, in order to moderate the effects (call this "Mitigation"). Third, one might take no action on the cause and so allow the effects to be realized at their full strength (call this "Acceptance"). Now consider four basic strategies for dealing with the effects of an impending change. First, one might try to eliminate the effect by taking pre-emptive evasive action (call this "Avoidance"). For example, if one is expecting a large sea level rise in the twenty-second century, one might prohibit new building on the coastline during the twenty-first century. Second, one might put in place a plan for evading damage when the effect arises (call this "Preparation"). So, for example, one might establish an infrastructure capable of responding very rapidly to extreme weather events. Third, one may simply count on one's ability to manage any adverse event if and when it occurs (call this "Coping"). For example, one may assume that one's existing capacities for dealing with other kinds of problems, such as the general emergency service infrastructure, will be sufficient to the task. Fourth, one may acknowledge that existing systems are not up to the task, but be resigned to taking whatever happens as it comes: i.e. one might decide to "weather the storm" (call this strategy "Endurance"). In this case, one may have defined other priorities, such as poverty and hunger, as so pressing that one cannot devote present resources to evading damages. These strategies are summarized in Table 8.2.

The core issue regarding the global test is whether institutions and theories prove themselves incapable (or perhaps simply unlikely) of responding appropriately to specific kinds of change by choosing a reasonable strategy (or set of strategies). For

Table 8.2 *Strategies for dealing with change*

	Response to cause	Response to effect	Implications for negative impacts
Prevention	Eliminate	- - - - - - -	Do not arise
Mitigation	Reduce	- - - - - - -	Moderated
Acceptance	Ignore	- - - - - - -	Full strength
Avoidance	- - - - - - -	Pre-emptive evasive action	Do not arise
Preparation	- - - - - - -	Plan for evasive action when effect arises	Moderated
Coping	- - - - - - -	Assume evasive action when effect arises	Moderated
Endurance	- - - - - - -	Absorb the costs	Full strength

example, it seems reasonable to describe the current global situation with respect to climate change as a combination of Acceptance and Endurance.[9] If so, and if a strong case could be made of there being a realistic threat of a Hard or Crash Landing, and that this makes the Acceptance and Endurance strategies unreasonable, then this would count as a criticism of the existing global system, and a failure of the global test.

Note that we need not assume that any particular combination of strategies, such as Acceptance and Endurance, is always unreasonable. The core issue with respect to the global test is whether existing institutions and theories are capable of choosing whatever strategy is reasonable for cases of particular kinds. However, there will be something suspicious about systems that endorse only one strategy very generally, i.e. as appropriate in a very wide variety of cases. This worry does arise about the existing system with respect to Acceptance, Endurance, and their close neighbors.

Suppose then that the situation is such as suggested above. In other words, in the case of climate change:

1. There is a realistic threat of a Hard or Crash Landing.
2. The current global situation is best described as manifesting strategies of "Acceptance and Endurance."
3. "Acceptance and Endurance" strategies are a product of the existing global system.
4. The nature of the threat makes "Acceptance and Endurance" strategies unreasonable.

Under such circumstances, there is strong reason to believe that the existing system is failing the global test. Here I shall not try to offer a comprehensive argument for (1), (2), and (3). Instead, I shall simply offer a few considerations that suggest that they are prima facie plausible.

---

[9] These options are perhaps too limited. Catriona McKinnon suggests to me that "deny and ignore" may be a more appropriate description of the recent global response. To my mind, "exacerbate and obstruct" also has its merits.

First, regarding (1), the possibility of the Hard and Crash Landing scenarios seems real enough, at least if one takes the perspective of several centuries. Observe, for example, that the IPCC's projections for temperature rise by 2100 under the more fossil fuel intensive ("business as usual") emissions scenarios is a best estimate of 3.4–4.0°C (likely range of 2.0–6.4°C) above the 1980–1999 average, and 3.9–4.5°C (likely range of 2.5–6.9°C) above the 1850–1899 average.[10] This is a very serious change. For comparison, the difference in global average temperature between now and the last ice age is roughly 5°C (though, of course, in the other direction), and that the last time the Earth experienced such high concentrations of carbon dioxide was 50 million years ago, during a period when crocodiles could be found at the poles. These facts prompt some scientists to say that the kind of change being projected would bring us essentially to a "different planet" than the one on which human civilization has evolved. Moreover, this change would occur very fast by geological standards – over one or two centuries, rather than many hundreds of centuries. Under such conditions, Hard and Crash Landing scenarios start to look plausible.

Second, regarding (2), the description of the recent (1990–2008) global strategy as one of "Acceptance and Endurance" seems reasonable. During that time, progress on mitigation has been extremely small. Instead of stabilization or reduction, global emissions have risen dramatically, as have emissions in almost all major countries. Global emissions are up by more than 30%,[11] and emissions from the United States, for example, are up more than 15%.[12] Moreover, there has been no substantial progress on adaptation, and indeed efforts in this direction have been substantially thwarted by the richer nations.

Third, regarding (3) there seems little doubt that this strategy for addressing climate change has emerged from current global institutions. Several attempts have been made to craft a better international response, but none have succeeded. In the end, Gore's pessimism has proven prescient. Now, some would object to both this and the second claim on the grounds that there is an impressive system of global governance in place in the Kyoto Protocol, and that this represents "by far the strongest environmental treaty that's ever been drafted" (David D. Doniger, Director of Climate Programs for the Natural Resources Defense Council, quoted in Brown, 2001). But I believe that this response is far too complacent. There are two basic reasons.

The first reason is that there can be no serious dispute about the fact that, considered as a global strategy, Kyoto has been a substantial failure. For one thing, the global rise in emissions is, on any account, huge, and cannot be ignored.

---

[10] Scenarios A2 and A1F1. The preceding temperature rises are against a baseline of 1980–1999. If one takes a baseline of 1850–1899, an extra 0.5 of a degree is added. (IPCC, 2007: 7).

[11] Global emissions were up by nearly 29.5% from 1990–2005 (Marland *et al.*, 2008), and emissions grew at a more rapid rate in 2007 (Moore, 2008).

[12] The numbers are against the baseline of 1990, rather than projected emissions. But the numbers for projected emissions are hardly more encouraging, since emissions are now at the high end of the IPCC's 1990 projections.

For another, if anything, the underlying trends are upwards, not downwards. Even after this surge, we do not seem to be curtailing humanity's growing appetite for fossil fuels. Realistic assessments project more and more growth in the future. Finally, even the more aggressive targets for future emissions reductions currently being touted – 20% reductions by 2020 – will only bring many countries back down to the levels of 1990. Hence, in effect, they simply offset the increases of the last 20 years. Given this, it is difficult to regard the policy of those years as any kind of "success."

The second reason is that the basic response offered to such complaints by enthusiasts for Kyoto – that Kyoto is only a first step in an evolving process, and we should not expect too much too soon – is unconvincing. I have addressed this argument in some detail elsewhere (Gardiner, 2004b), but it may be helpful to signal the general tenor of that response here. First, Kyoto is actually not the first step, but the outcome of nearly a decade of procrastination and false promises through the 1990s, followed by almost another decade of unhelpful distraction from the real task of bringing down emissions. Second, Kyoto has structural features that make it much less productive than might have been the case, and which will need to be overcome in future agreements. Thus, in some ways the Kyoto framework poses an obstacle to further progress. Third, those who regard Kyoto as a *necessary* first step implicitly endorse a very strong pessimism about what might have been possible between 1990 and 2012.[13] Yet I see no reason to accept such pessimism. Indeed, I suspect that endorsing it comes very close to conceding that not only the actual, but also any realistic alternative system of global governance must fail the global test. But these are extreme claims that ought not be conceded without argument.

Kyoto, then, does not cast doubt on the idea that global institutions have pursued a strategy of "Acceptance and Endurance." Indeed, one might go further and say that the sad history of the Kyoto process counts as one major contribution to the currently emerging failure of the global test. Despite the noble efforts of some, the dispute about Kyoto became just one facet of a general strategy of procrastination and delay at the global level, and the illusion of substantial action that it created often served as a distraction that facilitated these things.

Theoretical vices

Suppose that "Acceptance and Endurance" is unacceptable, and that this shows that existing institutions fail the global test. What might this reveal about contemporary political philosophy? Does this also fail? This question turns out to be more difficult to answer than one might think. The first complication is the general one that the

[13] For more on the dispute about Kyoto, see Gardiner (2004b) and DeSombre (2004).

connection between theories and institutions is likely to be imperfect at best. Given this, the worry arises that one cannot infer much about theories from institutional failure. Fortunately, in the present case, this concern does not seem too serious. Initially, there is at least some plausibility to the claim that current political institutions are, by and large, supported by the mainstream political theories (such as economic utilitarianism, libertarianism, Rawlsian liberalism, and cultural nationalism) or, more accurately, by some combination thereof, and that these theories themselves are often reflective of, and generated in response to, those institutions. More importantly, it seems unlikely that a closer correspondence between theory and practice will make a radical difference. Concern about our political theories is not merely derivative from worries about current institutions. Instead, the general imperviousness of most such theories to environmental issues, and to the concerns of the global test more generally, give us independent reasons to be troubled.

The second complication is that a theory might fall afoul of the global test in a variety of ways. For example, it might simply be silent on some important global threat, and so *oblivious*. But it may also encourage inaction, or else impede or block specific solutions, so that it is *complicit* in failure. Finally, a theory might preclude success altogether, and so *guarantee* disaster.

Unfortunately, such complaints have at least some initial credibility. In particular, much contemporary political theory does seem to have the effect of prioritizing other political concerns over those connected with the global test. For one thing, it has, until very recently, been focused on the individual and state level, largely neglecting global and intergenerational concerns. This supports the charge of obliviousness. For another, current work tends to concentrate on institutions that emphasize the short-term, local and national aspects of political affairs, such as democratic elections on three- to six-year cycles, market mechanisms, and the rights of current individuals. Thus, it is not crazy to think that it may be complicit in, or even go someway towards generating, global failure.

The third complication is that the assessment of rival political theories does not occur in a neutral evaluative setting. Recall that in the perfect moral storm theoretical inadequacies are of special interest because our choice of political theory might itself be corrupt. For example, if the intergenerational dimension – the fact that one generation can benefit from activities that pass serious costs on to its successors – dominates motivation, then we might expect earlier generations to prefer political philosophies that facilitate such buck-passing. In such situations, where the temptation to moral corruption is high, we must take extra care that our evaluation of theories is not distorted. One concern, of course, is that we will praise the wrong approaches. But another is that we will be too forgiving of error. For instance, in normal contexts obliviousness often seems a less serious shortcoming than other causes of failure. But in the perfect moral storm, silence may be a fatal flaw. Consider, for example, future generations. Obliviousness to their concerns should not be taken lightly, since it may disguise a morally unacceptable indifference to the future, or a worrying blindness to one of the central concerns of the subject. For comparison, what would we think of a political theory that placed a (perhaps impressive) account of intellectual property at its center, but had little or nothing to say about basic rights and political legitimacy, or one that was obsessed with etiquette but silent on everything else? Such myopia would surely be criticized, and for good reason. So, why be indulgent of political theories that are largely mute on the issue of the global test? The worrying answer is that it is because they address our concerns, and leave aside those that we would rather not see addressed.

The fourth complication is the difficulty of successfully accusing contemporary political philosophy of *anything* in particular (call this "the Teflon Problem") In particular, it is possible to characterize most theories at a very high level of abstraction, and at such dizzy heights most theories are so drained of content that they verge on vacuity. Suppose, for example, that one says that utilitarianism is ultimately about "bringing about the best," or that Kantianism is about "respecting" persons or treating them "as ends," or that rights-based theories are ultimately about "protecting the individual." At these levels of description, the content of each view is radically underdetermined. But this suggests that charges such as "utilitarianism fails the global test" will always be met with derision, especially by partisans. Surely, the thought goes, there is some (perhaps hitherto unimagined) version that will do the trick!

Given the Teflon Problem, it is tempting to retreat to claims like the following: theories of general type X *in their current or dominant manifestations* are incapable of dealing with climate change. But should we retreat in this way? Such limited claims would be interesting in their own right, and might be sufficient for many purposes. So, we should not denigrate them. Still, they can seem a little weak. In particular, they invite the following objection: if all that is being said really is that approach X hasn't got it right yet, how interesting (ultimately) is that charge? Can't we just say that we already know that our theories are imperfect, and that all the criticism really amounts to is "try harder"?[14]

This last complication makes it tempting to give up on deploying the global test against theories. Perhaps the claim of failure is just too difficult to prosecute, and the payoff of such prosecution too elusive, to be worth the trouble. This temptation is powerful. Still, I believe that we should resist it. First, there is simply too much at stake. The concern raised by Houghton's complaint, and highlighted by the global test, is just too central to concede this easily. Indeed, ignoring it seems to amount to a serious abdication of theoretical responsibility. Second, in any case, the emphasis on successful *prosecution* of claims of failure seems misguided. Presumably, the main

[14] I am grateful to Justin D'Arms for discussion on this issue.

point of introducing the test is not to convict any particular political philosophy, but rather to provoke a more general shift in focus. After all, we are much less interested in scoring partisan points, than in engaging with the problem, and with the general project of doing moral and political philosophy. In short, if the global test provides a genuine condition of adequacy for political theory, then fair-minded philosophers of all camps will want to take it seriously and try to make progress with it. In that case, we need not focus on successful prosecution as such: for example, on efforts to generate and then apply a set of necessary and sufficient conditions for inadequacy, or to pin down the criticism decisively for all comers, including the zealots. Instead, it will be enough merely to show that there is genuine cause for concern, and for this we might be satisfied with lower standards of proof. For example, just as in civil, as opposed to criminal, trials, we might accept a preponderance of the evidence approach, rather than insisting that the existence of a problem be shown beyond any reasonable doubt before we can proceed. After all, given that the stakes are so high, the former seems more than sufficient to justify further investigation.

Let us return then to the Teflon Problem. How are we to react to claims, such as that a given theory must *somehow* be able to deal with climate change; that we already know that our theories are imperfect; and that all the global test amounts to is an exhortation to try harder? An obvious initial worry is that global failure is a serious matter, so that the response seems more than a little glib (for example, think of how we might react to the proponent of an etiquette-centered theory of morality who made the same claims about his neglect of basic human rights). In addition, some ways of "not getting it right yet" are surely suspicious. For instance, we would have good reason to be skeptical of any approach that claimed that it could *always* adapt itself to any "new" set of concerns, however distant from its traditional ones.

To elaborate on this thought, let us consider some circumstances under which too much malleability seems to be a bad thing, revealing a flaw or vice of a particular approach.[15] One ground for suspicion arises if a theory turns out to be *unduly reactive*: it can mold itself to whatever trouble comes from the world or from other theories, but that trouble has to come first. In the face of something as severe as a potential failure of the global test, being reactive in this sense seems to make a theory overly *complacent*. A second, related flaw arises when an approach appears initially blind to concerns that are, or ought to be, morally fundamental. Both Houghton's claim and the global test suggest that some considerations have a certain kind of priority over others, and we might expect a political theory to wear such concerns on its sleeve, rather than discover them "late in the game" in response to a specific threat. An approach that is initially blind in this way appears to be guilty of a

---

[15] In invoking "vice," I mean merely to signal that in their exiting forms the approaches display a contingent but stable negative disposition.

troublesome *opacity* (and perhaps also *obliviousness*). Third, and more generally, if a theory turns out to be extremely malleable, we might wonder about its internal integrity. Whilst it is true that we do not want our theories to be inflexible and dogmatic in the face of new information and unexpected challenges, complete malleability would also be a problem. For one thing, infinitely pliable theories run the risk of becoming *vacuous*, functioning only as convenient labels for whatever happens to be on our minds at the time. For another, even if it does not lead to vacuity, excessive malleability threatens to make theories too *evasive*. We expect political theories to play a role in guiding action and justifying institutions. If they are to do this effectively, then they must already (explicitly or implicitly) address the major challenges we face.

An illustration: utilitarianism

If an approach to moral and political theory is oblivious, complacent, opaque, or evasive, then these are significant objections to it. Let us briefly illustrate and explore such concerns by focusing on a particular kind of moral and political theory, utilitarianism.[16] Generally speaking, utilitarianism holds that "we are morally required to act in such a way as to produce the best outcomes," where outcomes are usually evaluated in terms of human welfare (Jamieson, 2007: 164).[17] Hence, as a distinctively political doctrine, it claims that social and political institutions should be arranged towards the same end. This is an attractive view, and has been deeply influential in philosophy, economics, and law for several centuries. In his excellent recent paper, Dale Jamieson advocates a utilitarian approach to the global environmental crisis, in general, and climate change, in particular. In doing so, he emphasizes an attraction that is of special interest to us:

[U]tilitarianism has an important strength that is often ignored by its critics: it requires us to do what is best. *This is why any objection that reduces to the claim that utilitarianism requires us to do what is not best, or even good, cannot be successful.* Any act or policy that produces less than optimal consequences fails to satisfy the principle of utility. Any theory that commands us to perform such acts cannot be utilitarian.

(Jamieson, 2007: 164; emphasis added)

[16] I emphasize at the outset that the point of this discussion is merely to illustrate and explore. In particular, the point is not to put forward a comprehensive or decisive objection to utilitarianism; indeed, though I do not take myself to be a utilitarian, I suspect that suitably sophisticated versions of the view probably escape the charges made below. In particular, I am sympathetic to Dale Jamieson's worry about the gap between conventional categorizations of utiltiarianism and the views of its most illustrious defenders (see Jamieson, 2007: 169).

[17] Jamieson does not include an appeal to welfare as part of his definition, but his subsequent remarks are otherwise in sympathy with it. Of course, many utilitarians, including Jamieson, would extend concern to nonhuman animals as well. However, it seems fair to say that such considerations are not normally at the forefront of utilitarian political theory, and indeed may pose a major challenge to such theory, as usually conceived. Hence, I leave that complication aside here.

In short, Jamieson asserts that utilitarianism is invulnerable to a certain kind of objection: if a theory leads to worse outcomes, then it cannot be utilitarian. Moreover, in explaining this claim, he emphasizes the extreme malleability of the approach: "Utilitarianism is a universal emulator: it implies that we should lie, cheat, steal, even appropriate Aristotle, when that is what brings about the best outcomes" (Jamieson, 2007: 182).

Let us call this claim that a theory that leads to worse outcomes can't be utilitarian, "Jamieson's dictum." The dictum makes utilitarianism look good in the face of the global test, since it suggests that one virtue of the approach is that it cannot lead us to disasters like Crash Landing.[18] More generally, the dictum resonates with an important truth that matters to both utilitarians and most non-utilitarians: specifically, that the consequences of our behavior are extremely important, perhaps in some circumstances overridingly so.

There is an obvious sense in which Jamieson's dictum must be correct. If one takes utilitarianism as a thesis about the ultimate justification of social and political systems, then there are clear ways in which a genuinely utilitarian global system *could not* fail the global test. Still, Jamieson's emphasis on malleability should give us pause. It suggests that this defense of utilitarianism comes at a price. Given our discussion above, the appeal to malleability threatens to make utilitarianism an extremely *complacent* and *evasive* approach to political theory. The trouble arises because, even if we are secure in our knowledge that a global system that severely failed the global test could not *in the end* be a good utilitarian system, this information alone does not provide us with any guidance. In particular, we are no further along in knowing whether any particular system *currently* being advocated as utilitarian really is one. Utilitarianism becomes bulletproof, but only at the cost of *opacity*.

Let me illustrate this worry though a brief discussion of actual utilitarian thinking in climate change policy and more generally. As we shall see, utilitarianism can be cashed out in a number of different ways. However, the most influential version with respect to climate change has been the use of cost–benefit analysis (CBA) within a conventional economic framework. CBA is a tool of project evaluation that claims that the projects that should be pursued are those that give the maximum net benefits. Hence, economists try to calculate the benefits and costs of various policies for addressing climate change, and claim that the best policy is the one that maximizes net benefits.

---

[18] There are complications, of course. As usually understood, utilitarianism claims that the right thing to do is to maximize happiness. But this doctrine may lead us to some outcomes that other moralists would be inclined to view as disasters. For example, in principle, the view may sanction massive rights violations for the sake of greater happiness, it may justify the otherwise premature extinction of humanity if the benefits to the present are high enough, or it may lead to what Derek Parfit has called the Repugnant Conclusion (Parfit, 1986). But I leave aside these wider issues here. Given that Jamieson's definition of utilitarianism leaves the notion of "best outcome" opaque, he is not vulnerable to such worries.

This approach quickly raises some of the concerns listed above.[19] Consider first *opacity*. Different economic assessments of climate change come up with very different answers. One reason is that projecting costs and benefits into the long-term future is a difficult, if not impossible, task. How are we to know precisely what the global economy will look like in 50 or 100 years time, given that we do not know exactly which technological and social changes will occur, and what the specific negative effects of climate change will be (Broome, 1992: 10–11; Stern, 2008)? This problem is so severe that John Broome once claimed that CBA for climate change "would simply be self-deception" (Broome, 1992). In the perfect moral storm, this is a worrying thought. Still, the main point here is simply that, even if in principle CBA could tell us what we should do, the correct CBA for climate change is necessarily inaccessible to us at this point. In short, appeals to Jamieson's dictum are of no help for the decisions that need to be made.

Second, consider *complacency*. Here the prime suspect is the standard way in which CBA deals with future generations.[20] For one thing, economists typically assume that future generations will be richer than we are. But this assumption is threatened by the Hard Landing and Crash Landing scenarios. More generally, in conventional CBA the benefits and costs that accrue to future people are subject to a positive social discount rate. This means both that they count as less simply because they are in the future, and also (because of compounding) that impacts in the further future are worth dramatically less at current prices than current effects. On the face of it, this is a highly questionable and poorly justified practice that heavily favors the interests of current people.[21] Hence, there are real worries about moral corruption.

Third, CBA is prone to *vacuity* and *evasiveness*. Since there are no remotely secure numbers for either future costs and benefits or the social discount rate, the approach is extremely malleable, and in a way which threatens its internal integrity. As the economist Clive Spash puts it:

[E]conomic assessment fails to provide an answer as to what should be done. The costs of reducing CO_2 emissions may be quite high or there may be net gains *depending on the options chosen by the analyst*. The benefits of reducing emissions are beyond economists' ability to estimate so the extent to which control options should be adopted, on efficiency grounds alone, is unknown.

(Spash, 2002: 178)[22]

This gives rise to the worry that a suitably motivated economist could essentially justify whatever result she or he wanted. Given the temptation of moral corruption, this is a disturbing state of affairs.

---

[19] Jamieson (1992) raises similar criticisms.
[20] CBA also has trouble dealing with the value of nature. See, for example, Sagoff (1988).
[21] This is controversial. For defenses of such claims, see Cowen and Parfit (1992) and Gardiner (2006b).
[22] See also Azar and Lindgren, 2003: 253.

CBA also faces a deeper, and less often noticed, problem: it is not obviously the best way to implement utilitarianism. Worse, there are strong reasons to think otherwise. It is well known in utilitarian circles that calculating the net benefits of courses of action on each occasion is often a very poor way of maximizing total benefits. There are a number of reasons for this.[23] But the crucial point for our purposes is simply that *it is far from clear that either utilitarians, or even those with other moral views who share a concern for maximizing benefits, should support CBA.* In my view, it is hard to overstate the importance of this problem. Taken seriously, it threatens to undercut the basic rationale for the whole approach. At a bare minimum, it implies that the claim that CBA is a good method for maximizing net benefits ought not simply to be asserted or accepted without argument.[24]

The deep problem suggests a more general vice of the utilitarian approach, which emerges from the following story. There are many versions of utilitarianism, and CBA is most closely related to act-utilitarianism, the doctrine that one should aim to maximize the net benefits of each of one's actions. In the recent history of moral philosophy, act-utilitarianism has been subject to two major objections. The first to emerge was the complaint that utilitarianism neglects the individual. In focusing on the total happiness, it was said, utilitarianism puts no weight on how happiness is distributed. This may lead to the violation of what we usually think of as individual rights, and also to highly unequal distributions. Utilitarians responded to this objection in a number of ways. Some simply denied that rights or equality are important moral and political values. But most tried to diffuse such concerns by arguing that respecting individual rights and promoting equality usually contributes to greater happiness, and so these concerns should be offered special protection on utilitarian grounds. In particular, in response to the objection, many utilitarians gave up act-utilitarianism and came to advocate "rule-utilitarianism," the doctrine that the right thing to do is to act in accordance with the set of social rules which would maximize happiness.

A second standard objection to utilitarianism emerged later. It claimed that both act- and rule-utilitarianism neglect the role of individual agency in morality. Hence, for example, moral philosopher Bernard Williams complained that utilitarians are committed to seeing agents as completely in the service of the impersonal demands of maximizing happiness, and so do not account for the role of the agent's own values and personal attachments in moral action (see Smart and Williams, 1973). In response, many utilitarians argued that they could accommodate such concerns by

---

[23] One is that it is often impossible to predict the specific features of the future with any degree of confidence; another is that making calculations may itself involve high costs; a third is that acting on calculations may undermine other social goods, such as personal relationships and bonds of community. Other reasons also arise.

[24] This challenge should not be surprising. For one thing, it is just the flip-side of Jamieson's claim about malleability. For another, independent evidence that conventional CBA must face such scrutiny comes from many of its (officially nonutilitarian) critics. They often seem to be arguing that CBA causes more harm than good (or at least than some alternative policy).

focusing on the character traits and relationships characteristic of good utilitarian actors. In particular, some came to endorse an approach called "character (or virtue) utilitarianism," the doctrine that the right thing to do is to develop the set of character traits most conducive to maximizing happiness.

The point of this (no doubt simplistic) story is this. The shift in focus from acts to rules to characters raises a worry mentioned earlier. If utilitarianism merely reforms itself in response to any serious objection – molding itself to whatever trouble comes from the world or from other theories, but only when that trouble comes first – then it seems *unduly reactive*. This threatens its ability to play one of the main roles we might expect of a political theory, that of guiding us towards good social systems. If the approach is also oblivious, opaque, and evasive, this worry becomes even more serious.

In short, even if Jamieson's dictum is, strictly speaking, correct – a theory which claimed to be utilitarian but leads us to catastrophe could not be the correct utilitarian theory – this obscures an important consideration. If standard utilitarian thinking leads us to catastrophe, then it will be cold comfort to the survivors to be told that, by the standards of Platonic heaven, it could not have been utilitarian after all. From the point of view of the global test, the questions that really matter are whether *we* – those who have to make decisions about climate change and other global environmental problems – should be utilitarians in our actions, policies, and institutions, or whether utilitarianism can tell us what we should be.[25] But the answers to these questions remain unclear.[26] Given this, standard utilitarian thinking (such as CBA) might well fail the global test. To continue to endorse it merely because of Jamieson's dictum would be a very dangerous form of complacency indeed.

The upshot of this discussion is that, Jamieson's dictum not withstanding, the utilitarian approach is vulnerable to the vices identified above, and so might fail the global test. This is so despite the illusion of invulnerability bought through an appeal to abstraction. More generally, if the global test constitutes a genuine condition of adequacy, then fair-minded theorists of all camps will want to take such vices seriously and seek to address them.

Before closing, I want to be clear about the importance of the above argument. First, I do not mean to single out utilitarianism, as such, for criticism. Clearly, proponents of many other political theories will be tempted to say that a global system that results in catastrophe cannot be good by their lights because its effects on their favored set of concerns – e.g. human rights, property rights, communities, etc. – are extremely negative. The point I'm making is that there is something

---

[25] Traditional debates over whether utilitarianism can function as an esoteric doctrine, or is self-effacing or self-defeating, lurk in the background here. I cannot take on these questions here; but I do not believe that the current point rests on an unduly controversial position on those issues.

[26] Jamieson, of course, ultimately argues on utilitarian grounds that *we* – the ones having to act – should be virtue theorists. Hence, as I say below, his own view is not vulnerable to this objection.

genuinely suspicious about *all* such responses, and so we ought to expect more from our theories than this. Second, I do not take myself to be offering a decisive objection either to utilitarianism or to those other theories (of the sort just mentioned) that share some consequentialist concerns. My concern is with attempts to dismiss criticisms of existing approaches based on the global test by appealing to their most abstract versions. My complaint is that such appeals are vulnerable to important objections that can become especially serious in a context where global failure is possible and moral corruption likely. Complacency, evasiveness, and opacity are serious vices for a political theory to have.[27]

Human security

In this chapter, I have proposed a global test for social and political institutions and theories, and suggested that current varieties of both appear to be failing that test. I have also disputed the claim that some theories do not fail, because they might, or even must, in principle be able to solve the test. Against this, I claimed that theories can fall short in other ways, such as by being overly oblivious, complacent, opaque, and evasive. Moreover, I argued that such vices are both more likely and more damning in the presence of a perfect moral storm where there is serious risk of moral corruption.

How then might we move forward? One promising avenue would be to shift the focus away from the prevailing economic paradigm for evaluating institutions. Instead of worrying about how to maximize or optimize overall benefits, understood in market terms, the core concern might be with securing central goods, such as human rights, basic needs, and capabilities. Such a shift is already an important part of the debate about climate change and global environmental issues more generally, and is reflected in the emergence of concepts such as sustainability and human security.

Indeed, approaches that focus on human security at the individual level and the institutions needed to ensure it may have significant advantages here. On the one hand, they aim to deliver something that many around the globe do not yet have, and that Hard Landing and Crash Landing scenarios threaten for most current and future

---

[27] In addition, my quarrel is not with Jamieson himself. For one thing, I have admitted that considered as a thesis about ultimate justification, Jamieson's dictum must be correct: utilitarianism is, ultimately, bulletproof. What I would take issue with is the claim that this allows utilitarianism to escape the global test for political theories. In addition, I do not think that Jamieson's own utilitarian theory is vulnerable to these objections: it is not complacent, evasive, or opaque. Jamieson advocates that individuals cultivate a demanding set of green virtues that are not contingent on the behavior of others. The problem for him is whether he can show that such virtues are really justified on utilitarian grounds. But this is to take on the problem of malleability, not to avoid it. As it happens, it is not clear that Jamieson succeeds in this task. Walter Sinnott-Armstrong, another utilitarian, has recently argued for the contrary claim: that individuals should not be blamed even for engaging in self-indulgent environmentally destructive behavior – such as driving big SUVs just for fun. On his view, the appropriate obligations are at the political, not individual, level. This disagreement between Jamieson and Sinnott-Armstrong naturally raises worries about opacity and evasiveness. See Sinnott-Armstrong (2005).

people.[28] On the other hand, they can appeal to theorists of many different types, whether their core convictions concern rights, justice, or even utility. As some evidence for the latter, we might note what John Stuart Mill, the great utilitarian theorist, says about the "extraordinarily important and impressive kind of utility" connected with the related issue of physical security:

[S]ecurity no human being can possibly do without ... Our notion ... of the claim we have on our fellow creatures to join in *making safe for us the very groundwork of our existence* gathers feelings around it so much more intense than those concerned in any of the more common cases of utility that the difference in degree ... becomes a real difference in kind.
(Mill, 1863: Chapter V; emphasis added)

In conclusion, I have argued that climate change should be of serious concern to political philosophy. So far, this challenge has been largely ignored, in both academia and the public realm. Instead, scientific, economic, and short-term geopolitical discussions fill the journals, newspapers, and airwaves. In the abstract, this is puzzling. How could we be so oblivious and complacent in the face of such a potentially catastrophic threat? Unfortunately, the perfect moral storm offers an unflattering answer to this question. We need to wake up to that fact if we are to pass the global test.

Acknowledgements

This chapter aims to motivate a set of concerns about the political philosophy of climate change for an interdisciplinary audience. I defer deeper philosophical analysis for another occasion. Early versions of this material were presented at conferences at the University of Bremen and the University of Reading, and also to the fellows' seminar at the Center for Human Values at Princeton University. I am grateful to those audiences, and especially to Justin D'Arms, Nir Eyal, Axel Gosseries, Dale Jamieson, Stephen Macedo, Catriona McKinnon, John Meyer, Lukas Meyer, Cara Nine, and Philip Pettit. I also thank Karen O'Brien and Berit Kristoffersen for their encouragement. I am especially indebted to Asunción St. Clair and Lynn Gardiner.

References

Azar, C and Lindgren, K. 2003. Catastrophic events and stochastic cost–benefit analysis of climate change. *Climatic Change*, **56**(3), 245–55.

Broome, J. 1992. *Counting the Cost of Global Warming*. Isle of Harris, UK: White Horse Press.

Brown, P. 2001. World deal on climate isolates US. *Manchester Guardian*, July 24, 2001.

Cowen, T. and Parfit, D. 1992. Against the social discount rate. In P. Laslett and J. S. Fishkin, eds., *Justice Between Age Groups and Generations*. New Haven, CT: Yale, pp. 144–61.

[28] Moreover, they suggest that other concerns might be less fundamental, and so more tractable.

DeSombre, E. 2004. Global warming: more common than tragic. *Ethics and International Affairs*, **18**, 41–6.

Dryzek, J. 1987. *Rational Ecology: Environment and Political Economy*. Oxford, UK: Blackwell.

Gardiner, S. 2004a. Ethics and global climate change. *Ethics*, **114**, 555–600.

Gardiner, S. 2004b. The global warming tragedy and the dangerous illusion of the Kyoto Protocol. *Ethics and International Affairs*, **18**(1), 23–39.

Gardiner, S. 2006a. A perfect moral storm: climate change, intergenerational ethics and the problem of moral corruption. *Environmental Values*, **15**, 397–413.

Gardiner, S. 2006b. *Why do Future Generations Need Protection?* Working Paper. Chaire developpement durable. Available online: http://ceco.polytechnique.fr/CDD/PDF/DDX-06–16.pdf.

Gardiner, S. 2009. Saved by disaster? Abrupt climate change, political inertia and the possibility of an intergenerational arms race. *Journal of Social Philosophy*, **40**(2), 140–62. Special Issue on Global Environmental Issues, edited by Tim Hayward.

Gardiner, S. In press. *A Perfect Moral Storm: Climate Change: Intergenerational Ethics and the Global Environmental Tragedy*. New York, NY: Oxford University Press.

Houghton, J. 2003. Global warming is now a weapon of mass destruction. *Manchester Guardian*, July 28, 2003.

IPCC 2007. *Climate Change 2007: Synthesis Report*. Cambridge, UK: Cambridge University Press.

Jamieson, D. 1992. Ethics, public policy and global warming. *Science, Technology, Human Values*, **17**(2), 139–53.

Jamieson, D. 2007. When utilitarians should be virtue theorists. *Utilitas*, **19**(2), 160–83.

McKibben, B. 1989. *The End of Nature*. New York, NY: Random House.

McKibben, B. 2001. Some like it hot: Bush in the greenhouse. *New York Review of Books*, July 5, 2001.

Marland, G., Boden, T. and Andreas, R. J. 2008. *Global CO2 Emissions from Fossil-Fuel Burning, Cement Manufacture, and Gas Flaring: 1751–2005*. Carbon Dioxide Information Analysis Center, United States Department of Energy. Available online: http://cdiac.ornl.gov/trends/emis/glo.htm.

Mill, J. S. 1863. *Utilitarianism*. London: Parker, Son and Bourn.

Moore, F. 2008. *Carbon dioxide emissions accelerating rapidly*. Earth Policy Institute. Available online: http://www.earth-policy.org/Indicators/CO2/2008.htm.

Parfit, D. 1986. *Reasons and Persons*. Oxford, UK: Oxford University Press.

Sagoff, M. 1988. *The Economy of the Earth*. Cambridge, UK: Cambridge University Press.

Sinnott-Armstrong, W. 2005. It's not my fault. In W. Sinnott-Armstrong and R. Howarth, eds., *Perspectives on Climate Change*. Oxford, UK: Elsevier, pp. 221–53.

Smart, J. J. C. and Williams, B. 1973. *Utilitarianism: For and Against*. Cambridge, UK: Cambridge University Press.

Spash, C. L. 2002. *Greenhouse Economics: Value and Ethics*. London: Routledge.

Stern, N. 2008. The Economics of Climate Change. *American Economic Review*, **98**(2), 1–37.

Part IV

Reflexivity

9

Linking sustainable development with climate change adaptation and mitigation

LIVIA BIZIKOVA, SARAH BURCH, STEWART COHEN AND
JOHN ROBINSON

Introduction

Climate change impacts, and potential adaptive and mitigative responses, have been the subject of major assessments by the Intergovernmental Panel on Climate Change (IPCC), including the Fourth Assessment Report, which was published in 2007. Throughout the assessment process, increasing attention was focused on linkages between climate change responses and sustainable development, in part because climate change adds to the list of stressors that challenge the ability to achieve the ecologic, economic and social objectives that define sustainable development. Development choices, furthermore, can inadvertently result in altered vulnerabilities to climate variability and change, changed patterns of energy and material consumption and, consequently, emissions of carbon dioxide and other pollutants. Risks to human security could increase because global climate change interacts with specific regional stresses, including ecosystem degradation, economic difficulties, exposure to climate-related impacts, low response capacities and weak governance systems at sub-national and national scales (Barnett and Adger, 2007).

Klein *et al.* (2005) suggest that climate policy can evolve to facilitate the successful embedding of climate change within broader development goals to help reduce vulnerability and insecurity, but the integration of adaptation and mitigation at different operational scales remains a challenge (Jones *et al.*, 2007). Although these responses are widely regarded as complements rather than substitutes, gaps in our understanding of the various capacities that are required to carry out these responses have prevented a truly integrated assessment of response options.

In this chapter, we explore in more detail a possible methodology for linking sustainable development (S), climate change adaptation (A) and mitigation (M), herein referred to as 'SAM'.[1] We seek an approach that enables the explicit

---

[1] This chapter builds on a series of case studies published in a special issue of *Climate Policy* edited by Bizikova *et al.* (2007).

Climate Change, Ethics and Human Security, eds. Karen O'Brien, Asunción Lera St.Clair and Berit Kristoffersen.
Published by Cambridge University Press. © Cambridge University Press 2010.

consideration of climate change as part of the search for development paths that achieve the three pillars of economic, environmental and social sustainability in a particular local context. The specific objectives for conducting such local studies include the following:

1. To explore ways of transitioning to sustainable futures at the local level that anticipate mitigation and adaptation needs;
2. To assess ways of strengthening necessary capacities for effective responses to climate change that can be fabricated into development activities that promote win–win policy solutions, while addressing trade-offs;
3. To explore opportunities for the engagement of local stakeholders in a way that fosters collaboration, encourages creative thinking and promotes shared learning in addressing future development challenges;
4. To provide long-term guidance for local policies by strengthening the linkages between current local situations and future development options in the context of climate change impacts.

This chapter introduces a novel conceptual framework for SAM studies, describing key components of the proposed assessment framework. It then outlines key elements of a methodological approach for conducting SAM case studies, broadly characterised as a participatory integrated assessment (PIA). The proposed methodology incorporates a merging of model-based and participatory approaches for information gathering, analysis and communication, as part of a shared-learning experience, engaging researchers and stakeholders. We propose that an integration of the concept of 'capacity to respond to climate change' and an explicit consideration of barriers to responses will add value to previous participatory integrated assessment approaches to climate change.

Conceptual framework

As noted in Robinson *et al*. (2006), two ways to think about the linkages between adaptation, mitigation and sustainable development are to view sustainability as a possible consequence of climate policies (seeing sustainability through a climate change lens) and to view climate change mitigation and adaptation as rooted in, and the consequence of, different socioeconomic and technological development paths (seeing climate change through a sustainability lens). Given these two approaches, Bizikova *et al*. (2007) have proposed a two pronged or 'combined lens' approach to SAM, in which climate change and sustainable development goals are explicitly articulated (Figure 9.1) and simultaneously considered. This means that climate change responses become part of a portfolio of measures that represent new and more sustainable development pathways. This could include, for instance, specific actions designed to reduce consumption of fossil fuels and to avoid high intensity

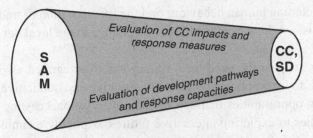

Figure 9.1 Climate change and sustainable development through a SAM lens. This 'combined lens' builds on the climate change (CC) and sustainable development (SD) lenses from Robinson *et al.* (2006).

development in vulnerable areas such as high-risk zones for flooding or drought. In practice, however, it has not been easy to 'mainstream' climate change measures into broader development decisions (e.g. Beg *et al.*, 2002; Agrawala, 2005; Schipper and Pelling, 2006). Consequently, a SAM assessment must consider potential barriers, constraints and tradeoffs that could affect the implementation of such measures. We suggest that many of these barriers are deeply rooted in path-dependent development trajectories, which have, in this context, been given scant analytical and theoretical attention in the past.

The combined SAM lens represents an acknowledgement that entry points are needed in order for development paths and climate change measures to be linked in the assessment process (see Figure 9.1). Development paths are created within a societal context that varies for each location, and give rise to the pools of resources or capacity that are available to be utilised in response to risks such as climate change. Climate change response measures, however, have at times been portrayed as inhibitors of the development aspirations of certain regions and countries. In the process of the design and implementation of the Kyoto Protocol, the concern has arisen that emissions reduction would be harmful to economic growth, or that various unequal commitments would create an unfair advantage for those countries that have negotiated relatively easily achieved emission reduction targets or none at all (Shimada, 2004). These arguments reveal the importance of wisely selected, contextually appropriate, response measures that take into consideration locally significant development priorities.

The search for entry points for the SAM assessment requires identification of key variables that can influence the results of both quantitative analyses and dialogue. For example, does the structure of a land-use model include explicit usage of a climate parameter (temperature, precipitation, etc.) or of a variable that can be derived from climate information (e.g. water supply, crop growth, forest pest risk, malaria risk) so that climate change can be factored into decision making? Similarly, if a climate-impacts model does not include parameters that represent adaptation capabilities, including human behaviour and the incorporation of tradeoffs between alternative decisions, how can the effects of a change in the local development path be evaluated?

These issues suggest the need for a framework of integrated assessment that is flexible enough to incorporate a range of quantitative and qualitative inputs, builds on the learning opportunities that can be generated by backcasting and scenario-based approaches to exploring alternative futures, explicitly examines the various capacities required for climate change responses and investigates the path-dependent institutional, technological and sociocultural barriers to effective responses.

Linkages to capacity

Recent discussions of the potential implications of climate change have led researchers to consider the resources and tools that provide the foundations upon which climate change responses are built. The concepts of mitigative and adaptive capacity were introduced in the Third Assessment Report of the Intergovernmental Panel on Climate Change (McCarthy *et al.*, 2001; Metz *et al.*, 2001) and further developed in the recent Fourth Assessment Report (Metz *et al.*, 2007; Parry *et al.*, 2007). Adaptive capacity, or the potential or ability of a system, region or community to adapt to the effects or impacts of climate change (Schneider *et al.*, 2001), is argued to be determined by factors such as the range of technological options, the availability and distribution of resources, the structure of critical institutions and the stocks of human and social capital (Yohe and Tol, 2002). Similarly, mitigative capacity represents the ability of a group to 'reduce anthropogenic emissions of greenhouse gases or enhance natural sinks', (Winkler *et al.*, 2006) and has, in the past, consisted of a set of determinants that are virtually identical to those of adaptive capacity (Yohe, 2001).

Since the development of these concepts, it has been noted that many of the proposed determinants of capacity are simply features of a highly developed, often industrialised nation that is rich in all forms of capital and possesses highly complex institutions (Burch and Robinson, 2007). Thus, we see that many of the determinants of mitigative capacity are in fact part of some broader pool of resources that can be utilised in response to a multitude of risks, and that are closely linked to the underlying development path of a nation or community. This broader pool of resources has been called 'response capacity' and represents the human ability to respond to any risk with which it is faced (Burch and Robinson, 2007), including the management of greenhouse gases and the consequences of their production (Tompkins and Adger, 2005). Mitigative and adaptive capacities, therefore, are better thought of as the institutions and policies (derived from the underlying response capacity and thus a group's level of development) which are geared specifically towards the mitigation of and/or adaptation to climate change. For example, the creation of a government agency aimed at managing climate change adaptation, or passage of an energy efficiency policy, represents the conversion of generalised response capacity into, respectively, adaptive and mitigative capacity.

These concepts provide insights into the ways in which responses to climate change are rooted in the underlying development path of a group or nation, by way of response capacity. Furthermore, institutional, technological and sociocultural barriers to effective climate change action, which grow out of complex path dependent processes, are revealed. These barriers may inhibit the translation of capacity into action on climate change and as such require special attention throughout the PIA process. Incorporating capacity into this analysis is crucial for two reasons: it reveals the resources with which any response to climate change can be built and it draws attention to the underlying development path, which simultaneously influences both capacity and barriers to action.

Assessment framework

Although the ways in which participatory processes are useful in integrated assessments are manifold, one especially relevant use is the capacity to frame problems and support the policy process by designing and facilitating policy debate and argumentation (Hisschemöller *et al.*, 2001). In simple terms, an assessment of a complex problem like climate change can involve researchers and stakeholders. If an 'integrator' can bring together the suppliers of the science information with those who are demanding a particular kind of information, then between the two groups it may be possible to obtain what Rotmans (1998) calls an integrative narrative, which helps to define the problem, leading to a consensus building process, and a sense of joint ownership in the process. This relationship becomes more complex with the involvement of sponsors of research and independent organisations, as well as the interested public and researchers mentioned above. The knowledge that emerges from these exchanges creates what has been called 'interactive social science' (Caswill and Shove, 2000). The following sections introduce a framework with which interactive social science can be operationalised in the context of SAM.

Participatory Integrated Assessment

As a complement or alternative to all-inclusive integrated models, the participatory integrated assessment (PIA) approach is a framework that utilises dialogue as a research tool. PIA is an umbrella term describing approaches in which non-researchers play an active role in integrated assessment (van Asselt and Rijkens-Klomp,

2002), and can be used to facilitate the integration of biophysical and socioeconomic aspects of climate change adaptation and development (Hisschemöller *et al.*, 2001). Van Asselt and Rijkens-Klomp (2002) identify several approaches, including methods for mapping out diversity of opinion (e.g. focus groups, participatory modelling) and reaching consensus (e.g. citizens' juries, participatory planning). Huitema *et al.* (2004) have reported on a recent exercise on water policy that employed citizen's juries. PIA has also been used to facilitate the development of integrated models (e.g. Turnpenny *et al.*, 2004) and to use models to facilitate policy dialogue (e.g. van de Kerkhof, 2004). PIA has evolved in part from Participatory Action Research (PAR), which is a well-known approach that social scientists have used in studies of traditional practices and environmental knowledge of aboriginal communities (see, for example, Krupnik and Jolly, 2002; Reid *et al.*, 2006).

In a PIA, individuals agree to participate in a process that allows them to use dialogue to approach a complex problem. Throughout the PIA process, issues often arise, and must be overcome, when bringing together people with potentially disparate points of view on an issue. Confrontations or contradictory information may also arise through this process, which requires reconciliation. The dialogue is intended to find a way to navigate through these issues without the process taking on the atmosphere of a judicial inquiry or other form of legal proceeding. Dialogue thus provides the 'scaffolding' with which participants can relate new experiences to existing knowledge (Chermack and van der Merwe, 2003). The purpose of this dialogue is not simply outreach, nor even simply one-way teaching. Instead, this is intended to be two-way or multi-voice teaching. A PIA can create a shared learning experience for scientists, business interests, community representatives, aboriginal peoples, resource managers, governments or any stakeholder with knowledge to share and a reason to be part of the process (Figure 9.2). In other words, the knowledge that is created during the process of PIA is an emergent property of the interactions among multiple actors (Robinson and Tansey, 2006), thus the use of participatory process in integrated assessment can be seen as a 'learning machine' rather than a 'truth machine' (Berkhout *et al.*, 2002).

Dialogue and models can be mutually reinforcing (Tansey *et al.*, 2002), and together, can improve decision making by integrating knowledge from a variety of sources (Hisschemöller *et al.*, 2001). For instance, dialogue can support model building through the process of participatory modelling, including, for example, mediated modelling (van den Belt 2004; Robinson and Tansey, 2006). Still, the success of a PIA is heavily dependent on the presence of the following elements: sufficient time, subject matter of the dialogue that is relevant to the participants and the explicit presentation of uncertainty and disagreement (Hisschemöller *et al.*, 2001). Time is especially important, as evidenced by the Netherlands' COOL project

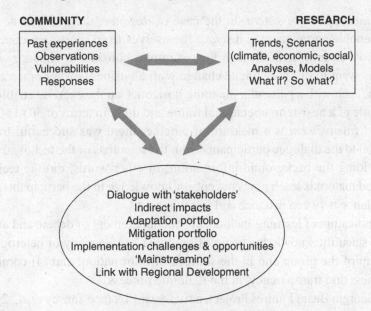

COMMUNITY

Past experiences
Observations
Vulnerabilities
Responses

RESEARCH

Trends, Scenarios
(climate, economic, social)
Analyses, Models
What if? So what?

Dialogue with 'stakeholders'
Indirect impacts
Adaptation portfolio
Mitigation portfolio
Implementation challenges & opportunities
'Mainstreaming'
Link with Regional Development

Figure 9.2 Framework for SAM shared learning involving researchers and communities of interest (based on Cohen and Waddell, 2009).

(Climate OptiOns for the Long term), which focused on Dutch national policy for reducing emissions and extended over several years (van de Kerkhof, 2004).

Part of what distinguishes PIA from the traditional construction of integrated models is the application of dialogue techniques. Dürrenberger *et al.* (1997) provide a categorisation of various dialogue techniques according to two criteria:

1. Embeddedness, which is related to the level of activity within a decision-making process, ranging from low (information gathering) to medium (advice) to high (decision); and
2. Level of conflict, which ranges from absent to latent to acute.

In a situation with high embeddedness and acute conflict, the technique most likely to be used is mediation. In a research situation with low to medium embeddedness, there are other options, such as focus groups, planning cells and consensus conferences. Climate change research, dialogue or negotiation can include situations that cover much of this range of embeddedness and conflict. But, the choice of dialogue exercise really depends on what the objective is. For example, policy exercises and focus groups are techniques that are designed to bring out the range of positions, rather than to force a consensus. In other words, in these exercises, the task here is not to reach agreement, but rather to find out what all the positions really are.

Returning to the example of the COOL project, van de Kerkhof (2004) explored a number of dialogue exercises and attempted to measure how well these different exercises worked. This was determined through several indicators. Two of these were *distance* and *involvement*. In the case of *distance*, the question is: does the approach enable participants to distance themselves from short-term concerns and focus on wider long-term issues? For example, consider a situation in which researchers want to discuss climate change with a business owner or a manager of a reservoir, each with a particular planning horizon. Can the exercise enable them to think outside of a near-term operational frame and think in terms of 30 to 50 years in the future? *Involvement* is a measure of whether there was successful transfer of information to the dialogue participants from the scientists or the technical staff that were providing the background information. In other words, did the technical or background materials teach new concepts or knowledge to the participants? Also, is there a balance between distance and involvement?

Other indicators of learning include: (1) encouragement of debate and argument; (2) use of scientific knowledge being offered; (3) homogeneity or heterogeneity in the makeup of the group and in the sources of information; and (4) commitment, trust, fairness and transparency, in the dialogue process.

In the Georgia Basin Futures Project, a five-year PIA (see Tansey *et al.*, 2002), the focus was explicitly on the co-production of knowledge, whereby 'expert' knowledge was combined with partner knowledge at multiple stages of the project in order to give rise to an emergent understanding of sustainability options at a regional scale. The focus was much less on the communication of technical knowledge to stakeholders than on the co-production of understanding about the choices and consequences facing the region (Tansey *et al.*, 2002; Robinson *et al.*, 2006). Such work brings forth complex question about power, trust, and the nature and status of different forms of understanding (Robinson and Tansey, 2006).

Models as dialogue starters

Success in a PIA or any dialogue process will ultimately depend on convincing stakeholders to agree to remain committed to the process. In that respect, it will likely be more difficult to organise and sustain dialogue within a PIA than for a modelling group to construct an integrated model on its own, because response rates can be influenced by many external factors, including other commitments of participants (e.g. to their jobs and families). At the same time, however, PIA can include group-based model construction, in which stakeholders contribute directly to model construction, within a modelling process that includes the incremental development of codes and functions, tested and evaluated by local practitioners and other local knowledge holders (van den Belt, 2004). This offers an exercise in shared learning, which is important for providing a sense of ownership in the process, as well as the results (Rotmans and van Asselt, 1996). Furthermore, the output of the process is fundamentally a product of the involvement of 'users'

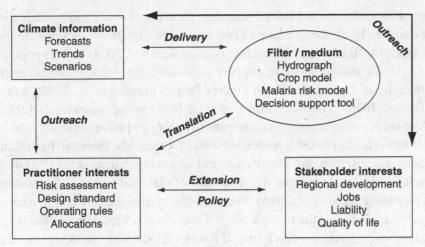

Figure 9.3 Role of models as 'filters' to translate basic climate information into indicators of interest to practitioners and stakeholders (adapted from Cohen and Wadell, 2009).

of the research, rather than strictly the result of design and execution by researchers in isolation from other actors.

What then is the role of models (including group-based models) in initiating and maintaining a dialogue between researchers and stakeholders? In the context of climate variability and change, one important function is the translation of information from one base of knowledge (such as scientific information or traditional environmental knowledge) into other forms of knowledge (see Figure 9.3). The role of local professionals and technical support staff (i.e. practitioners, such as engineers and resource managers working for local/regional governments) is an important element of this translation process for decision makers.

The long-term sustainability of dialogue processes is critical to the success of participatory approaches. Models can play an important role as dialogue starters, and can offer interactive learning opportunities. For such processes to be successful as shared learning experiences, they have to be inclusive and transparent. Haas (2004) describes examples of experiences in social learning on sustainable development and climate change, noting the importance of sustaining the learning process over the long term, and maintaining distance between science and policy while still promoting focused science-policy interactions. Applications of focus group and other techniques for stakeholder engagement are described for several studies in Europe (Welp *et al.*, 2006) and Africa (Conde and Lonsdale, 2004).

Group-based model building was used to study water resources in the Okanagan region of Canada, where a systems dynamics model of stocks and flows of water was used to explore various response options for adapting to climate change and population growth (Langsdale *et al.*, 2007). Other participatory examples are case studies of agriculture in the United Kingdom (Lorenzoni *et al.*, 2000), adaptation to worst-case sea level rise in Europe (Toth and Hizsnyik, 2008), options for greenhouse gas mitigation in the Netherlands (van de Kerkhof, 2004) and the application of the QUEST model series in support of stakeholder engagement on regional sustainability in the Georgia Basin Futures Project (Tansey *et al.*, 2002; Robinson and Tansey, 2006). The latter incorporates a backcasting approach (see below), which enables model users to explore pathways for producing desirable outcomes.

Despite the host of benefits associated with PIA methods, there can be difficulties in reaching consensus on identifying and engaging participants (Parkins and Mitchell, 2005), and in interpreting the results of dialogue within different communities (see Huntington *et al.*, 2006). There are also challenges inherent in measuring the quality of dialogue, the transparency of process, the promotion of learning and indicators of influence (Rowe and Frewer, 2000; van de Kerkhof, 2004). Furthermore, there is the danger that such processes will add to the stress already being felt by local-scale decision makers and institutions, as external pressures from national and international scales are downloaded onto their jurisdictions (Burton *et al.*, 2007). Allen (2006) notes that increased responsibility without increased capacity could create a barrier to successful participation in shared learning exercises on long-term climate and sustainability, which could indeed lead to the disempowerment of local participants. A related issue is the potential for claims of oppressive or patronising research, which may result from improperly acknowledging distributions of power (Orme, 2000; Cooke and Kothari, 2001).

Operationalising the assessment framework

In this section, we review the opportunities and challenges associated with scenarios and backcasting. The former offers a picture of changes evolving from current conditions to various future states. The latter involves a subset of scenarios which are the result of a process that begins with defining futures, ideal or otherwise, and then works backwards to current conditions.

Scenarios

Climate change is a long-term problem in which the past may not be the only or best guide to the future. Scenarios represent an excellent opportunity to begin an exploration of different futures in which climate can be treated as a variable condition, rather than as a constant state with regular oscillations. There is a growing interest in the use of scenarios as heuristic tools that make mental maps more explicit (Berkhout *et al.*, 2002), as aids to social and organisational learning (Chermack and van der Merwe, 2003), as tools for scanning the future in a rigorous, creative and policy-relevant way that explicitly incorporates normative elements (Swart *et al.*, 2004), and as a means to explore the effects of alternative course of action for future problems involving multiple actors, risk and uncertainty (Mayer *et al.*, 2004). The three most commonly used types of scenarios are exploratory scenarios, which posit a range of underlying socioeconomic conditions upon which alternative futures may be constructed; extrapolatory scenarios, which provide forecasts based on baseline trends; and normative scenarios, or backcasting, which are built on positive and negative visions of the future, and explore pathways of change that might lead to them (Berkhout *et al.*, 2002). This section will first consider the broad umbrella of scenarios in general, and will then focus on backcasting as a subset of scenarios that may be particularly useful for SAM analyses.

Assessments of future scenarios can lead to the realisation by researchers and local partners that current operational and planning practices may need to be re-examined, and current vulnerabilities reconsidered, as part of a larger process of defining and implementing local-scale sustainable development paths. A scenario-based SAM case study would ideally try to address three critical questions:

1. *What if?* In a scenario of climate change and development, what kinds of local impacts may occur? Without absolute certainty regarding future climate conditions, can a damage report be provided for various combinations of climate change and local development choices?
2. *So what?* Does the damage scenario make a difference? As the damage scenario is presented to interested parties, such as irrigation purveyors, municipal planners, business leaders or engineers, the dialogue can turn to whether the damage scenario makes a difference to their vision of the future. Current planning processes may consider population growth or changes in important industries or market conditions. The climate change impacts scenario represents a new set of climate statistics translated into a physical (and possibly an economic) impact. Could this scenario hinder long-term efforts to meet local development goals?
3. *What can be done?* How can a sustainable development pathway be defined for the study area? What adaptation measures should be considered? How could these become 'mainstreamed' into a sustainable development pathway? How could emission reduction measures become part of this without creating new vulnerabilities? If climate change scenarios can be translated into parameters that are relevant to stakeholders in their planning context, then it should be possible to start a dialogue about adaptation, mitigation and development that would be different from the initial planning scenario, yet still be plausible.

Ultimately, the goal is to move away from the simplifying assumption that everything else remains equal and towards the notion that 'the ground is moving under our feet, while the atmosphere is changing over our heads.' In order to do this, it is important to consider that a search for the most likely future, which is the most common approach of scenarios, may be misguided or counterproductive, since the future remains to be created (Höijer *et al.*, 2006). Furthermore, the goals of a SAM approach, as articulated above, include the integration of human responses to climate change within the broader context of sustainable development paths. Thus, it seeks to explore and articulate the path that a community or group wishes to take, rather than the path that is most likely. In doing so, a normative element is explicitly incorporated into the more traditional use of scenarios in Participatory Integrated Assessment. Backcasting is a scenario approach that attempts to accomplish this goal and will be briefly introduced in the following section.

Backcasting

Backcasting is a method of analysing alternative futures (Robinson, 1994; Dreborg, 1996). Unlike predictive forecasts, backcasts are not intended to reveal what the future will likely be, but to indicate the relative feasibility and implications of different policy goals. It is thus explicitly normative, involving working backwards from a particular desired future end point or set of goals to the present, in order to determine the physical feasibility of that future and the policy measures that would be required to reach it.

While the value and quality of a predictive forecast depend upon the degree to which it accurately suggests what is likely to happen under specified conditions, backcasting is intended to suggest the implications of different futures, chosen not on the basis of their likelihood but on the basis of other criteria defined externally to the analysis (e.g. criteria of social or environmental desirability). No estimate of likelihood is possible since such likelihood would depend upon whether the policy proposals resulting from the backcast were implemented. Thus, while the emphasis in forecasts is upon discovering the underlying structural features of the world that would cause the future to come about, the emphasis in backcasts is upon determining the freedom of action, in a policy sense, with respect to possible futures.

In order to undertake a backcasting analysis, future goals and objectives are defined and then used to develop a future scenario, analysing the technological and physical characteristics of a path that would lead toward the specified goals. The scenario is then evaluated in terms of its physical, technological and socioeconomic feasibility and policy implications. Iteration of the scenario is usually required in order to resolve physical inconsistencies and to mitigate adverse economic, social and environmental impacts that are revealed in the course of the analysis.

In what have been called first order backcasting methods, the general nature of the desired end point is specified by the research team in advance of the backcasting analysis itself, which focuses on the detailed characteristics of the end-point future and/or the path between that end point and the present. Second order backcasting is based on a view of backcasting as a social learning process, whereby not just the analysis of the feasibility of a scenario, but also the choice of the goals themselves, should be part of the analysis (Robinson, 2003). The decision as to what is a desirable scenario is thus an emergent property of the process of analysis. This in turn requires the development of iterative and participatory modelling tools and processes quite different from those usually used (Quist and Vergragt, 2006). The type of participatory integrated assessment described above, however, is well-suited to the goals of backcasting and the combination of these two approaches can yield creative, open and reasonable solutions to problems, while explicitly acknowledging the challenges presented by uncertainty, contradiction and ambivalence (Höijer *et al.*, 2006).

Since backcasting approaches explicitly introduce the question of policy choice, they serve to refocus the use of analysis away from responding to inevitable futures and toward exploring the nature and feasibility of alternative directions of policy. This helps to put the onus for choosing back where it belongs: in the policy arena. Furthermore, the explicitly normative approach built into backcasting methods parallels the norms and values that are embedded within an integrated SAM research agenda.

Linking climate change and sustainable development

Addressing such complex questions as sustainable development and climate change requires a coordinated effort building on linkages between research and practice. As stated earlier, the ultimate goal of a SAM case study is to assist in moving policy-making at the local level toward sustainability, given new challenges arising from changing climate. Information available from climate impact/adaptation assessments needs to be integrated with local development priorities, including mitigation of GHG emissions, in order to inform the policy process, take action and strengthen capacities. This aim can be accomplished by moving towards integrated assessments based on the interaction with stakeholders, enhancing interdisciplinary work and producing outcomes that can be included in the decision-making processes. The key elements for conducting the case study, which are discussed below, build upon PIA and utilise scenario-based models developed though public dialogue processes (van Asselt and Rijkens-Klomp, 2002).

Defining the preferred local sustainable development scenario

A scenario focused on describing a sustainable future for the location of interest, and in this way providing the wider context for climate change impacts, is a crucial element of any SAM case study. By creating a context for climate impacts, future climatic conditions are no longer imposed on present-day socioeconomic conditions, an approach taken by many climate impacts studies that appears insufficient. Scenarios can provide heuristics that enable policy makers to identify possible future vulnerabilities to climate change and to assess the capacity of future societies to adapt to its impacts – impacts that only have meaning in their social context (Berkhout *et al.*, 2002). Approaches focusing on anticipating future development should capture the long-term and dynamic nature of local goals and challenges arising from efforts to achieve sustainability.

At an operational level, planners are more concerned about local development questions than global-scale climate change (Gupta and van Asselt, 2006). Therefore, by creating the local scenario, the stakeholders have an opportunity to identify local development goals, such as large-scale reforestation, urban densification, expanding the transportation network, building new energy facilities, agricultural change or other unexplored options that they are concerned about, and if the goals are designed properly they may help to move to a sustainable future. These goals can be seen as local problems, addressing a single dimension of sustainability, but analysed under different time frames (Swart *et al.*, 2004).

Earlier in this chapter, we discussed second-order backcasting as a way to explore the nature and feasibility of alternative directions of achieving desired future development goals in collaboration with stakeholders. In the case of urban infrastructure development, Ruth and Coelho (2007) showed that stakeholders can be strongly biased towards preexisting notions of development, and consequently, the selected method needs to allow stakeholders to distance themselves from past and current trends that may restrict opportunities to move towards a more sustainable path. Using backcasting to generate scenarios presents an opportunity to challenge current development pathways.

Finally, the scenario evolution is complemented by the analysis of an identified set of indicators. It is important that these indicators include barriers to action on sustainable development, such as institutional capacity, social capital, technological path dependences and modes of environmental governance. In the realm of adaptation research, scholars have developed a number of indicators that are closely related to the original set of adaptive and mitigative capacity determinants developed by Yohe (2001). Especially useful as a starting point are baseline measures, such as the number of deaths incurred by a natural disaster, the number of people affected by the disaster, the amount of damage done, the extent of development on floodplains, coasts and other vulnerable areas (Yohe and Tol, 2002). This provides an idea of the vulnerability of a region in the event of future, climate-related disasters, and thus the adaptive response that is required. Other indicators of adaptive capacity, which was defined above to be much more specific than previous definitions of the concept, would include the number of institutions and policies created to deal with the impacts of climate change, and the perceived efficacy and feasibility of these policies. Shepherd *et al.* (2006) have identified specific components related to feasibility, such as institutional and technical enabling factors and barriers to implementation. Generally speaking, these indicators assist in an integrated assessment of climate change response options. Furthermore, both future scenarios and backcasting, discussed below, can utilise these indicators to formulate visions of the future that are desirable to PIA participants.

Applying PIA offers integration of the biophysical and socioeconomic aspects of development, by creating opportunities for shared experiences in learning, problem definition and design of potential solutions.[2] For issues such as climate change, science needs to open up to new ways of framing problems, and in such a process stakeholder dialogues can play a vital role (see Welp *et al.*, 2006). Researchers should carefully design the scenario development exercise by giving sufficient time for problem definition, knowledge-base development, building trust and gaining a shared appreciation of critical development questions articulated by the participants (Swart *et al.*, 2004).

Linking local development and climate change impacts

Scenarios can help local communities move towards a sustainable pathway at the early stage of planning, while addressing complexity and risk arising from climate change. Scenarios provide a context to identify information on broader social and environmental consequences of climate change (UKCIP, 2001), in contrast to often narrowly defined climate change impacts assessments.

In defining the measures required to achieve the identified 'future' scenario, here we focus on whether climate change makes a difference in being able to achieve this sustainable future (i.e. does climate change impose any additional technical, environmental or social constraints). Here, models/tools specifically developed to address climate-related concerns, such as water, forests, food production, fish, property exposure/risk, health, etc. can be applied. This may also include group-based or mediated model building of the STELLA tradition (van den Belt, 2004). This PIA approach has been used in a series of studies on climate change and water management in the Okanagan region of British Columbia, using a range of population growth scenarios (Cohen and Neale, 2006; Cohen *et al.*, 2006; Langsdale *et al.*, 2007). However, linkage with a sustainable development scenario has not yet been attempted.

Lorenzoni *et al.* (2000) show that stakeholders in farming communities in the United Kingdom were more concerned when climate change was presented as an increase in rainfall rather then changes in temperatures, due to the higher sensitivity

---

[2] Welp *et al.* (2006) suggested that for issues such as climate change, science needs to open up for new ways of framing problems and, in such a process, stakeholder dialogues play a vital role.

of agricultural production to rainfall then temperature. Collaborating with stakeholders in assessing potential climate change damage, as well as the performance of adaptation measures, is crucial for translating outcomes of models that are often presented as changes in temperature, precipitation or sea level rise into units that matter to local stakeholders. For example, changes in biomass accumulation, changes in length of the growing season or water availability can be seen as relevant information for stakeholders. Estimating the impacts of climate change is a task for researchers, applying their expertise within an interdisciplinary team and communicating the results to the stakeholders (for details see Figure 9.3). This provides a challenge for natural scientists dealing with downscaling climate predictions and linking them with changes in natural resources, but also creates opportunities for interdisciplinary scholars to deepen their understanding of social implications of climate change impacts.

Assessing the relationship between adaptation and mitigation at the local level

Viewing adaptation and mitigation responses as part of broader development strategies on the path towards sustainability offers a new way of thinking about climate change response and creates opportunities for innovative approaches. In a climate change context, for example, this would involve near-term objectives for both adaptation and mitigation alongside objectives that characterise an improved future capacity or ability to address adaptation and mitigation (Wilson and McDaniels, 2007). Adaptation options can include a diversity of measures, which can also strengthen response capacities. Examples include new regulations for infrastructure development, revised allocation principles for water, introduction of new agricultural crops or ways of production, changed zoning to avoid vulnerable sites, heat alert systems and new flood management systems. Similarly, a diversity of mitigation options can be promoted as feasible in a local context to reduce emissions such as renewable energy production, support for low emission and clean technologies, increasing energy efficiency, recycling and reuse, urban densification and zoning for multiple use, carbon sequestration and lowering emissions from agriculture.

Here, there are opportunities for scientists to articulate examples of adaptation and mitigation measures, but it also gives members of the collaborative group the opportunity to address their own experiences. The main focus should be on creating an inventory of specific adaptation and mitigation measures, including the assessment of performance through various indicators of development (e.g. vulnerability/risk exposure, economic performance and environmental indicators including GHG emission rates). Specific attention would be given to identifying measures with potential tradeoffs that we are trying to avoid, such as new emissions linked to adaptation measures and new vulnerabilities emerging from mitigation efforts.

Here, the dialogue should also include specific coverage of barriers to implementation (municipal policy, high-level policy and technical, economic and social/ethical concerns) or enabling factors based on current and potential changes in local response capacity. The identification of synergies between adaptation and mitigation helps to deepen the understanding of policy makers of how to promote efficient use of limited resources.

The inventory of measures should be drawn from diverse economic sectors involved in the local development scenario. For example, local transportation network development can minimise vulnerability from extreme weather events, but at the same time create opportunities for emission reduction and co-benefits (e.g. public transportation, car pools). Energy system development can strike a balance between increased use of renewable energy sources that might be affected by climate change (e.g. changes in precipitation may impact hydropower or climate impacts on forestry as a source of biomass) and decreasing dependency on coal and oil. Co-benefits such as less local air pollution and associated human health benefits and job opportunities should be considered. Similar examples can be recognised in the forestry sector, in which the selection of planted species needs to address potential changes in pest dispersion and occurrence of fires. Importantly, increased forest planting also provides other benefits such as flood protection, carbon sequestration, carbon stored in long-life wood products or the use of wood as a source of energy (Dang *et al.*, 2003; Swart *et al.*, 2003).

Local dialogue processes have the potential to identify synergies among local measures that could be utilised actively to mobilise local actors (Næss *et al.*, 2006). Deliberation during the PIA is focused on identifying potential synergies and trade-offs between adaptation and mitigation identified in the previous step, which requires an identification of capacities and actions, their cost and benefits for various sectors (e.g. diversification of energy sources, water resources management, forestry, agriculture and consequences for ecosystems) and their interaction in the development scenario. This creates a dialogue about the feasibility of the responses, how they can be incorporated into the development initiatives at the early stage of development planning. Participants would therefore gain a better appreciation of the need to balance competing priorities, preferences and decisions using limited resources.

Identifying measures to promote sustainable development pathways

The core of the SAM case study is the integration of local sustainable development goals and the climate change 'damages and opportunities' report. Long-term iterative scenarios would be the product of this integration. Similar to Næss *et al.* (2006), we see local problems as 'door openers' or entry points for the evolution of local climate policies and the possibility of creating mutual legitimisation within the mitigation and adaptation policy arenas. We should focus on informing community-level decision makers and other relevant stakeholders to allow them to adjust their long-term priorities to the threats (also opportunities) from a changing climate. Defining the local development scenario, with embodied climate change responses, can lead to the identification of a broader set of measures than targeting the responses solely as a mitigation and/or adaptation response (Swart *et al.*, 2004). This would require answering questions, such as: does the inclusion of identified adaptation and mitigation measures fit with local sustainability priorities, or create additional problems, requiring a second iteration? For example, do 'green' buildings require legislative protection of solar access?

Developing a local sustainable development scenario, identifying impacts of climate change and adaptation and mitigation measures are meant to facilitate 'learning by planning' within the group of practitioners by developing key elements of the SAM case study. Thus we focus on creating venues to operationalise the responses in the institutional context by identifying measures to promote 'learning by doing' at the policy levels (Wiek *et al.*, 2006). Examples with collaborative development of climate change impacts assessment showed that local decision makers involved during the whole assessment can internalise the results (UKCIP, 2001). Similarly, Moser (2005) showed that cases of successful linkages between information produced by scientists, and actually used in decision making, occurred in those cases in which ongoing interaction and mutual understanding of practitioner information needs vis-à-vis scientific capability was already further along. Consequently, the suggested outcomes can be internalised more easily if there is already an expressed need to address impacts of climate change based on past negative experiences such as floods (Penning-Rowsell *et al.*, 2006).

Although we emphasise the importance of the participatory process throughout the case study, it does not necessarily imply success in transforming the integrated SAM scenario to actual policies. However, the value of this stakeholder-driven approach goes beyond guiding further scientific inquiry. Such direct stakeholder engagement increases the likelihood that the decision makers will find subsequent research salient, credible and legitimate, insofar as the underlying assumptions are derived in part from their observations (Cash *et al.*, 2003). Moreover, this type of research product provides immediate educational benefits in a process of social learning for all participants, including researchers (Moser, 2005). Through the learning that will occur during the case study, we can create a collective process in which the policy makers, scientists and other stakeholders generate new insights into, and a better understanding of, the different perceptions, ideas, interests and considerations that exist with regard to the nature of the development goal in the context of climate change (van de Kerkhof and Wieczorek, 2005). This process will deepen their understanding of the appropriate strategies to induce the transition towards the preferred local scenario and to facilitate the changes identified in the scenario. Policy recommendations could include new rules and standards, building codes, revised principles of natural resources management and policy incentives for using new technologies. This type of collaboration between the stakeholders also creates important outcomes in the form of new relationships and social capital built among players who would not ordinarily interact, much less do so constructively (Hajer, 2005).

A SAM case study requires that stakeholders identify institutional constraints and promote institutional partnerships that can foster the implementation as well as the monitoring of identified responses. Innes and Booher (2004) showed that collaborative planning processes addressing developmental priorities are essential ways to build societal capacity and institutional capacity. It is important to create partnerships between institutions targeting climate change and sustainable development in order to establish a platform for the successful implementation of the outcomes of the case study.[3]

Finally, the outlined approach should be applicable in diverse local contexts with their own challenges. Local initiatives depend not only on the decisions made at the local level, but also respond to trends occurring and policies being adopted at the regional, national and international scale. Many responses to climate change, especially those focused on mitigation, are developed at the international level and then translated into commitments at national and regional levels. However, there are examples of local initiatives emerging from shared learning experiences which offer useful models for carrying out SAM studies with PIA, such as the guidebook for adaptation based on experiences in King County, Washington, in the United States (Snover *et al.*, 2007). Future initiatives need to focus on conducting local case studies, strengthening the transfer of lessons learned between cases and facilitating the collaborative work of diverse groups of stakeholders including local decision makers as well as participants from other levels of governance. These studies should also address technical and institutional aspects of adopting development decisions involving both climate change adaptation and mitigation, which could provide useful insights for development of an integrated global climate change adaptation and mitigation agenda.

Acknowledgements

We gratefully acknowledge the continuous support of the Adaptation and Impacts Research Section (formerly Division) of Environment Canada for the research involved in this chapter and for the SAM initiative.

---

[3] Examples from the UK show that creating local climate change impacts scenarios within the socioeconomic development path in collaboration with local development bodies promoted greater local ownership of regional policies and increased commitment to their implementations (UKCIP, 2001).

References

Agrawala, S., ed. 2005. *Bridge over Troubled Waters: Linking Climate Change and Development*. Paris: Organization for Economic Co-operation and Development Publishing.

Allen, K. M. 2006. Community-based disaster preparedness and climate adaptation: local capacity-building in the Philippines. *Disasters*, **30**, 81–101.

Barnett, J. and Adger, W. N. 2007. Climate change, human security and violent conflict. *Political Geography*, **26**, 639–55.

Beg, N., Corfee-Morlot, J., Davidson, O. *et al.* 2002. Linkages between climate change and sustainable development. *Climate Policy*, **2**, 129–44.

Berkhout, F., Hertin, J. and Jordan, A. 2002. Socio-economic futures in climate change impact assessment: using scenarios as learning machines. *Global Environmental Change*, **12**, 83–95.

Bizikova, L., Robinson, J. and Cohen, S., eds. 2007. Integrating climate change actions into local development. *Climate Policy*, Special Issue, **4**, 105.

Burch, S. and Robinson, J. 2007. Beyond capacity: a framework for explaining the links between capacity and action in response to global climate change. *Climate Policy*, **4**, 304–16.

Burton, I., Bizikova L., Dickinson, T. and Howard, Y. 2007. Upscaling the results of local research on adaptation, mitigation and sustainable development. *Climate Policy*, **4**, 353–70.

Cash, D. W., Clark, W. C., Alcock, F. *et al.* 2003. Knowledge systems for sustainable development. *PNAS* **100**, 8086–91. Available online: http://www.pnas.org/content/100/14/8086.full?sid=65702fcc-255f-4f93-a872-5845f44f4492.

Caswill, C. and Shove, E. 2000. Introducing interactive social science. *Science and Public Policy*, **27**, 154–7.

Chermack, T. J. and van der Merwe, L. 2003. The role of constructivist learning in scenario planning, *Futures*, **35**, 445–60.

Cohen, S. and Neale, T., eds. 2006. *Participatory Integrated Assessment of Water Management and Climate Change in the Okanagan Basin, British Columbia*. Final report, Project A846, submitted to Natural Resources Canada, Ottawa, Environment Canada and University of British Columbia, Vancouver.

Cohen, S. and Waddell, M. 2009. *Climate Change in the 21st Century*. Montreal: McGill-Queen's University Press.

Cohen, S., Neilsen, D., Smith, S. *et al.* 2006. Learning with local help: expanding the dialogue on climate change and water management in the Okanagan Region, British Columbia, Canada. *Climatic Change*, **75**, 331–58.

Conde, C. and Lonsdale, K. 2004. Engaging stakeholders in the adaptation process. In B. Lim and E. Spanger-Siegfried, eds., *Adaptation Policy Frameworks for Climate Change: Developing Strategies, Policies and Measures*. Cambridge, UK: Cambridge University Press, pp. 47–66.

Cooke, B. and Kothari, U., eds. 2001. *Participation: The New Tyranny?* London: Zed Books.

Dang, H. H., Michaelowa, A. and Tuan, D. D. 2003. Synergy of adaptation and mitigation strategies in the context of sustainable development: the case of Vietnam. *Climate Policy*, **3**, S81–S96.

Dreborg, K. 1996. Essence of backcasting. *Futures*, **28**, 813–28.

Dürrenberger, D., Behringer, J., Dahinden, U. *et al.* 1997. *Focus Groups in Integrated Assessments: A Manual for a Participatory Tool*. ULYSSES working paper, WP-97-2, Darmstadt: Darmstadt University of Technology.

Gupta, J. and van Asselt, H. 2006. Helping operationalise Article 2: a transdisciplinary methodological tool for evaluating when climate change is dangerous. *Global Environmental Change*, **16**, 83–94.

Haas, P. M. 2004. When does power listen to truth? A constructivist approach to the policy process. *Journal of European Public Policy*, **11**, 569–92.

Hajer, A. M. 2005. Setting the stage, a dramaturgy of policy deliberation. *Administration & Society*, **36**, 624–47.

Hisschemöller, M., Tol, R. S. J. and Vellinga, P. 2001. The relevance of participatory approaches in integrated environmental assessment. *Integrated Assessment*, **2**, 57–72.

Höijer, B., Lidskog, R. and Uggla, Y. 2006. Facing dilemmas: sense-making and decision-making in late modernity. *Futures*, **38**, 350–66.

Huitema, D., van de Kerkhof, M., Terweij, R., Van Tilburg, M. and Winsemius, F. 2004. *Exploring the Future of the Ijsellmeer: Report of the River Dialogue Project on the Dutch Citizen's Juries*, Amsterdam: Institute for Environment Studies, Vrije Universitat, 94 pp.

Huntington, H. P., Trainor, S. F., Natcher, D. C. *et al.* 2006. The significance of context in community-based research: understanding discussions about wildfire in Huslia, Alaska. *Ecology and Society*, **11**, article 40. Available online: http://www.ecologyandsociety.org/vol11/iss1/art40/.

Innes, J. E. and Booher, E. D. 2004. Reframing public participation: strategies for the 21st century. *Planning Theory & Practice*, **5**, 419–36.

Jones, R. N., Dettmann, P. Park, G., Rogers, M. and White, T. 2007. The relationship between adaptation and mitigation in managing climate change risks: a regional response from North Central Victoria, Australia. *Mitigation and Adaptation Strategies for Global Change*, **12**, 685–712.

Klein, R. J. T., Schipper E. L. F. and Dessai, S. 2005. Integrating mitigation and adaptation into climate and development policy: three research questions. *Environmental Science & Policy*, **8**, 579–88.

Krupnik, I. and Jolly, D., eds. 2002. *The Earth is Faster Now: Indigenous Observations of Arctic Environmental Change*. Washington, DC: ARCUS.

Langsdale, S., Beall, A., Carmichael, J., Cohen, S. and Forster, C. 2007. An exploration of water resources futures under climate change using system dynamics modeling. *The Integrated Assessment Journal*, **7**(1), 51–79.

Lorenzoni I., Jordan, A., O'Riordan T., Turner, K. and Hulme M. 2000. A co-evolutionary approach to climate change impact assessment. Part II: a scenario-based case study in East Anglia (UK). *Global Environmental Change*, **10**, 145–55.

Mayer, I. S., Carton, L., de Jong, M., Leijten, M. and Dammers, E. 2004. Gaming the future of an urban network. *Futures*, **36**, 311–33.

McCarthy, J. J., Canziani, O. F., Leary, N., Dokken, D. and White, K., eds., 2001. *Climate Change 2001: Impacts, Vulnerability and Adaptation: Contribution of Working Group II to the Third Assessment Report of the Intergovernmental Panel on Climate Change (IPCC)*. Cambridge, UK: Cambridge University Press.

Metz, B., Davidson, O. R., Bosch, P. R., Dave R. and Meyer, L. A., eds., 2001. *Climate Change 2001. Mitigation: Contribution of Working Group III to the Third Assessment Report of the Intergovernmental Panel on Climate Change*. Cambridge, UK: Cambridge University Press.

Metz, B., Davidson, O. R., Bosch, P. R., Dave R. and Meyer, L. A., eds., 2007. *Climate Change 2007. Mitigation: Contribution of Working Group III to the Fourth Assessment Report of the Intergovernmental Panel on Climate Change*. Cambridge, UK: Cambridge University Press.

Moser, S. C. 2005. Impacts assessments and policy responses to sea-level rise in three US states: an exploration of human dimension uncertainties. *Global Environmental Change*, **15**, 353–69.

Næss L. O., Norland, I. T., Laffertyb, W. M. and Aall, C. 2006. Data and processes linking vulnerability assessment to adaptation decision-making on climate change in Norway. *Global Environmental Change*, **16**, 221–33.

Orme, J. 2000. Interactive social sciences: patronage or partnership? *Science and Public Policy*, **27**, 211–19.

Parkins, J. R. and Mitchell, R. E. 2005. Public participation as public debate: a deliberative turn in natural resource management. *Society & Natural Resources*, **18**, 529–40.

Parry, M. L., Canziani, O. ., Parry, Palutikof, J. P., van der Linden, P. J. and Hanson, C. E., eds. *Climate Change 2007: Impacts, Adaptation and Vulnerability. Contribution of Working Group II to the Fourth Assessment Report of the Intergovernmental Panel on Climate Change.* Cambridge, UK: Cambridge University Press.

Penning-Rowsell, E., Johnson, C. and Tunstall, S. 2006. 'Signals' from pre-crisis discourse: lessons from UK flooding for global environmental policy change? *Global Environmental Change*, **16**, 323–39.

Quist, J. and Vergragt, P. 2006. Past and future of backcasting: the shift to stakeholder participation and a proposal for a methodological framework. *Futures*, **38**, 1027–45.

Reid, W. V., Berkes, F., Wilbanks, T. and Capistrano, D., eds. 2006. *Bridging Scales and Knowledge Systems: Concepts and Applications in Ecosystem Assessment.* Washington, DC: Island Press.

Robinson, J. 1994. Backcasting. In R. Raehlke, ed., *The Encyclopedia of Conservation and Environmentalism.* New York, NY: Garland Publishing Inc.

Robinson, J. 2003. Future subjunctive: backcasting as social learning. *Futures*, **35**, 839–56.

Robinson, J. and Tansey, J. 2006. Co-production, emergent properties, and strong interactive social research: the Georgia Basin Futures Project. *Science and Public Policy*, **33**, 151–60.

Robinson, J., Bradley M., Busby, P. *et al.* 2006. Climate change and sustainable development: realizing the opportunity. *Ambio*, **35**, 2–9.

Rotmans, J. 1998. Methods for IA: the challenges and opportunities ahead. *Environmental Modeling and Assessment*, **3**, 155–79.

Rotmans, J. and van Asselt, M. 1996. Integrated assessment: a growing child on its way to maturity. *Climatic Change*, **34**, 327–36.

Rowe, G. and Frewer, L. J. 2000. Public participation methods: a framework for evaluation. *Science, Technology & Human Values*, **25**, 3–29.

Ruth, M. and Coelho, D. 2007. Managing the interrelations among urban infrastructure, population, and institutions. *Climate Policy*, **4**, 317–36.

Schipper, L. and Pelling, M. 2006. Disaster risk, climate change and international development: scope for, and challenges to, integration. *Disasters*, **30**, 19–38.

Schneider, S., Sarukhan, J., Adejuwon, J. *et al.* 2001. Overview of impacts, adaptation, and vulnerability to climate change Climate Change 2001: Impacts, Adaptation, and Vulnerability. In J. McCarthy, O. Canziani, N. Leary, D. Dokken and K. White, eds., *Climate Change 2001: Impacts, Adaptation and Vulnerability.* Cambridge, UK: Cambridge University Press, pp. 75–103.

Shepherd, P., Tansey J. and Dowlatabadi, H. 2006. Context matters: what shapes adaptation to water stress in the Okanagan. *Climatic Change*, **78**, 31–62.

Shimada, K. 2004. The legacy of the Kyoto Protocol: its role as the rulebook for an international climate framework. *International Review for Environmental Strategies*, **5**, 3–14.

Snover, A. K., Whitely Binder, L., Lopez, J. *et al.* 2007. *Preparing for Climate Change: A Guidebook for Local, Regional, and State Governments*. Oakland, CA: ICLEI (Local Governments for Sustainability), 172 pp.

Swart, R., Raskin P. and Robinson J. 2004. The problem of the future: sustainability science and scenario analyses. *Global Environmental Change*, **14**, 137–46.

Swart, R., Robinson, J. and Cohen, S. 2003. Climate change and sustainable development: expanding the options. *Climate Policy*, **3**, S19–S40.

Tansey, J., Carmichael, J., Van Wynsberghe, R. and Robinson, J. 2002. The future is not what it used to be: participatory integrated assessment in the Georgia Basin. *Global Environmental Change*, **12**, 97–104.

Tompkins, E. and Adger, W. N. 2005. Defining response capacity to enhance climate change policy. *Environmental Sciences and Policy*, **8**, 562–71.

Toth, F. L. and Hizsnyik, E. 2008. Managing the inconceivable: participatory assessments of impacts and responses to extreme climate change. *Climatic Change*, **91**, 81–101.

Turnpenny, J., Haxeltine, A. and O'Riordan, T. 2004. A scoping study of user needs for integrated assessment of climate change in the UK context: Part 1 of the development of an interactive integrated assessment process. *Integrated Assessment*, **4**, 283–300.

UK Climate Impacts Programme (UKCIP) 2001. Socio-economic scenarios for climate change impact assessment: a guide to their use in the UK Climate Impacts Programme. Oxford, UK: UKCIP.

van Asselt, M. B. A. and Rijkens-Klomp, N. 2002. A look in the mirror: reflection on participation in integrated assessment from a methodological perspective. *Global Environmental Change*, **12**, 167–84.

van de Kerkhof, M. and Wieczorek, A. 2005. Learning and stakeholder participation in transition processes towards sustainability: methodological considerations. *Technological Forecasting & Social Change*, **72**, 733–47.

van de Kerkhof, M. 2004. *Debating climate change: a study of stakeholder participation in an integrated assessment of long-term climate policy in the Netherlands*. Utrecht: Lemma Publishers.

van den Belt, M. 2004. *Mediated modeling: a system dynamics approach to environmental consensus building*. Washington, DC: Island Press.

Welp, M., de la Vega-Leinert, A. and Stoll-Kleeman, S. 2006. Science-based stakeholder dialogues: theories and tools. *Global Environmental Change*, **16**, 170–81.

Wiek A., Binder C. and Scholz, R. W, 2006. Functions of scenarios in transition processes. *Futures*, **38**, 740–66.

Wilson, C. and McDaniels, T. 2007. Linking climate change adaptation, mitigation and sustainable development with structured decision-making tools: the Gateway Program example, *Climate Policy*, **4**, 353–70.

Winkler, H., Baumert, K., Blanchard, O., Burch, S. and Robinson, J. 2006. What factors influence mitigative capacity? *Energy Policy*, **35**, 692–703.

Yohe, G. and Tol, R. 2002. Indicators for social and economic coping capacity: moving toward a working definition of adaptive capacity. *Global Environmental Change*, **12**, 25–40.

Yohe, G. W. 2001. 'Mitigative capacity: the mirror image of adaptive capacity on the emissions side. *Climatic Change*, **49**, 247–62.

10

Global poverty and climate change: towards the responsibility to protect

ASUNCIÓN LERA ST.CLAIR

> One of the ironies about this financial crisis is that it makes action on poverty look utterly achievable. It would cost $5 billion to save six million children's lives. World leaders could find 140 times that amount for the banking system in a week. How can they now tell us that action for the poorest on the planet is too expensive?
>
> *John Sentamu, Archbishop of York*[1]

Introduction

I will start this chapter by stating the obvious – that the regions, communities and individuals that are already being affected by climate change and will be hardest hit in the future are those with the least capacity to cope with the consequences, and those who have contributed least to the problem. This includes poor countries, poor people and marginalised and vulnerable individuals and groups in all societies, including those living in advanced economies in the West. What follows from this, however, are three important claims that form the basis of my argument in this chapter. First, I argue that the most efficient way to promote sustainable adaptation to unavoidable climate change is the immediate eradication of severe poverty, the building of solid welfare systems and social protection, and the minimisation of inequalities. This calls for radical changes in the theory and practice of poverty reduction and a shift from the emerging focus on market-based solutions to climate change. Second, I argue that one effective and fair framing of the relations between global poverty and climate change is around the concept 'the responsibility to protect'. This concept can be interpreted in two ways: (1) that the world community has the political responsibility to protect people from poverty; and (2) that the world community has the political responsibility to protect poor people from the negative

[1] The Archbishop of York 2008. Speech to The Worshipful Company of International Bankers Dinner. Wednesday 24 September 2008. Available online: http://www.archbishopofyork.org/1980.

Climate Change, Ethics and Human Security, eds. Karen O'Brien, Asunción Lera St.Clair and Berit Kristoffersen. Published by Cambridge University Press. © Cambridge University Press 2010.

effects of climate change. These two claims can be summed up in one: *the world community has the responsibility to protect human security*, which includes the security of both current and future generations of human beings. And third, I argue that efforts for eradicating severe poverty and a focus on responsibilities are not only justified because of the intrinsic value of the dignity of all human beings. They also have a fundamental instrumental value, in that they will help building a culture of solidarity and global social cohesion – necessary elements for reaching regional and global agreements on mitigation strategies, for fair adaptation policies and for avoiding dangerous and irreversible environmental changes.

None of these claims, however, have strong currency in current debates about climate change, poverty and responses to these enormous and complex challenges. Building on many of the points made in other chapters of this volume, I suggest that we must be extremely critical of current perspectives linking poverty and climate change. One way to move discussions about framing and ethics forward in relation to climate change is to focus on normative concepts, such as human rights and political responsibility. There is no need to wait for greater scientific certainty, more and better measurements of effects and feedbacks of climate change on the poor or increased mortality, morbidity and poverty statistics. Addressing climate change (whether through mitigation or adaptation) is undeniably about broad-based societal decision making. It is an issue that requires dealing with diverse values and norms, as well as value conflicts. A key challenge is to move beyond misleading ideas about 'expert knowledge' on poverty and to emphasise instead the human aspects, the role of values in helping us envision and deal with complex global problems in alternative ways. Another challenge is to question the ways in which climate change, development and poverty reduction strategies are linked by diverse actors, from the global to local level. The dominant expert discourse on poverty, as represented by the World Bank and economistic and quantitative perspectives, has been constructed as a technical question, mainly resolved by growth. We have consequently judged societies according to their consumption achievements, rather than by the way they have treated the most vulnerable members. Climate change is also increasingly constructed as a technical question. One of the fundamental tasks for research is to challenge these constructions, both discursively and politically. The combined framing of poverty and climate change as technical rather than human issues and the proposals of market-based 'solutions' may lead to major mistakes and negative social outcomes.

The point of departure for this chapter comes from insights from earlier work on development ethics, and on the relations between processes of knowledge production on climate change and poverty, and on ethnographic studies of knowledge production in relation to value-based ideas and human rights within global development institutions (St.Clair 2006a, 2006b, 2007; McNeill and St.Clair, 2009;

Gasper and St.Clair, 2010). The chapter also builds on premises outlined by scholars offering critical perspectives on global poverty and its relationship to climate change (Deacon, 2007; Bond *et al.*, 2008; Gough *et al.*, 2008; Harcourt, 2008; Lawson and St.Clair, 2009) and engages with selected views addressing the human dimensions of environmental change, social challenges and framing issues (Verweij and Thompson 2006; Adger *et al.*, 2009; Heltberg *et al.*, 2009; Hulme, 2009).

The first section of the chapter argues against naïve framings of poverty and development in relation to climate change; it criticises the already visible transformation of the relations between climate change and poverty as just one more technical problem to be addressed by economic policy and technical innovation, and as a job to be done mainly by the development aid bureaucracies, donors and development NGOs. The second section illustrates the importance of learning from history in challenging these dominant framings, and points to the negative consequences of delinking the environment from human action and its global consequences. The third and concluding section suggests that a way forward is the reframing of climate change and poverty jointly as matters of political responsibility. Whether at the local, national or global level, actors with power and those able to respond are those who must make the immediate efforts to address climate change in ways that first and foremost protect the poor and vulnerable.

Dominant perspectives on the relations between climate change, poverty and development

Dominant perspectives on climate change from the aid community argue for some sort of pro-poor adaptive development (OECD, 2002), treating climate change as a threat to development achievements without questioning the theories and practices related to poverty reduction. Dominant thinking does not question the extent to which dominant perspectives have actually succeeded in addressing severe poverty. The suffering of the poor has been constructed as a problem that is unrelated to inequality, political economy or to various kinds of social relations; also framed as unrelated to questions of solidarity and cooperation within and across countries. These dominant framings displace rather than address ethical questions and responsibilities. As a point of departure for a debate on the relations between climate change, development and poverty, this can be considered a 'dead end'. Or worse, it may prioritise adaptation to the powerful's way of life, deepening inequalities and poverty, increasing global unfairness and leading to greater violations of human rights.

Whose knowledge counts and why in framing climate change, development and poverty change, is one of the key issues that must be investigated. First, there is a dominant narrative about what type of problem climate change represents (Hulme, 2009). A review of reports of the Intergovernmental Panel of Climate Change (IPCC) suggests that scientists in this multilateral knowledge organisation have become more concerned with the relationship between poverty, development and climate change in recent years, slowly increasing the space dedicated to the negative impacts on poor regions and poor populations. However, this community of experts has done so by adopting the language of development agencies and the major players in development aid, in particular, the language of the World Bank and a selected group of elite economists. As Hayward and O'Brien (Chapter 11) point out, this leads to claims by United Nations Framework Convention on Climate Change (UNFCCC) officials that the World Bank may be an appropriate institution to implement any climate agreement in relation to development issues emerging from the 2009 Copenhagen meeting.

Most of the voices we hear – the dominant voices – point to the threat that climate change poses to increasing development and reducing poverty in the global South. Discussions focus on the notions of adaptation and mitigation in developing countries and less so on mitigation in developed countries and the questioning of conceptions of development. The dominant view is summarised in a report jointly written by many multilateral development agencies and published by OECD in 2002:

Experience suggests that the best way to address climate change impacts on the poor is by integrating adaptation responses into development planning. This is fundamental to achieve the Millennium Development Goals, including the over-arching goal of halving extreme poverty by 2015, and sustaining progress beyond 2015.

(OECD, 2002: v)

The fundamental issue presented in this report, and by others that have followed, is that climate change is a 'major threat' to development and to ongoing efforts to reduce poverty. And their fundamental policy prescription is to create tools for 'adaptation aid'. We have seen emerging, for example, funding for so-called 'clean' development through the Clean Development Mechanism, the Special Climate Change Fund (SCCF) and more recently National Adaptation Plans for Action (NAPAs). These are new policy tools rapidly spreading across the developing world. Most of these tools rely on market-driven proposals while they cannot compare to the funds available through the private sector-driven Global Environmental Facility (GEF).

In addition to adaptation aid, donors are becoming increasingly interested in environmental protection as a means for climate change mitigation (e.g. supporting efforts to maintain and protect natural forests as these are natural carbon sinks). For example, the United Nations Reduced Emissions from Deforestation and Forest Degradation (REDD) has enticed countries, such as Norway, to invest major amounts of their aid funds in climate change mitigation. We are also witnessing a rapid revival of old versions of technology transfer, in this case the transfer of green technologies to developing countries as part of aid packages. At hand here is an overarching concern for integrating climate change in the work of development aid; to add further aid to developing countries under the category of adaptation and to attempt to promote a type of 'clean' (sustainable) development, preferably if it is also conducive to mitigation efforts and means for offsetting the accumulated effects of the dirty development of advanced and industrialised countries.

Updated versions of sustainable development, however, are not directed to environmental sustainability or to ideas promoted by environmental justice movements across the world, but to what is usually called 'low-carbon growth' or 'carbon neutral development'. As such this goal depends on innovation of various forms in order to be achieved (for example, by promoting growth activities that offset the overall levels of non-captured carbon). Protecting rainforests in some parts of the world to counterbalance energy-intensive development in other parts could still count as carbon-free growth. The point is that many of these suggestions are not necessarily concerned with poverty and the poor, but rather they represent a recasting of an old pro-growth vision of progress and development into a new context.

A focus on carbon trading and carbon sequestration as key tools towards an equitable distribution of emissions is not the result of a broad-based debate with relevant stakeholders. In fact, many critics and climate justice activists consider these solutions as 'bogus', 'false' or 'far from genuine'. Arguably, they are solutions about offsetting climate change as it affects unsustainable development patterns of the global North, and not necessarily the appropriate tools to address the enormous twin challenges of climate change and poverty reduction. To put it very clearly, climate change actors and experts and climate change negotiations are focused on climate change, not on poverty reduction; and the emergent discourse joining the problems takes for granted that we are already doing our best to tackle poverty. For some, this is considered a new form of 'carbon colonialism' that permits and facilitates the continuance of modernity and its privilege to the few because resources are locked away for the host countries. As Bond *et al.* (2008) rightly argue, the most destructive effect of this focus on carbon is that it allows us to believe we can carry on polluting.

The language of 'pro-poor adaptation' is thus about reducing the impacts of climate change, rather than questioning existing unfairness and misguided models of development. Attempting to introduce adaptation to climate change as one of the fundamental tasks of development has been an impressively fast-growing approach. A statement of progress issued by the OECD (2008) claims that there has been major progress in integrating climate change into development work. Commitments include, for example, assessments of climate vulnerability in ongoing projects, encouraging integration of NAPAs with Poverty Reduction Strategy Papers (PRSPs); and the development of tools and methodologies to assess climate vulnerabilities (OECD, 2008). Donors are laying out strategic aid programmes where climate change is at the core of their documents. For example, the Norwegian White

Paper 'Climate, Conflict and Capital' (NOU, 2009), identifies climate change as the most important challenge to overcome if the fight against poverty is to succeed:

Climate change is making the existing obstacles to eradicating poverty even greater. Without an international effort to assist developing countries in adapting to climate change, global poverty will increase. This is primarily a matter of reducing poor countries vulnerability to the impacts of climate change. Norway's efforts in the fight against poverty will also seek to address the problems caused by climate change. In order to be robust, environmental policy in both rich and poor countries must be linked to an economic policy that promotes employment and growth in income and production.

(NOU, 2009)

The UK's latest White Paper on International Development, *Building our Common Future* (DFID, 2009) similarly puts central weight on issues of growth, conflict and climate change, yet falls short of questioning the system that may reasonably be seen as having contributed to the climate crisis. Economics is arguably the dominating expert voice in policy debates about the 'human dimensions' of climate change. Elite voices such as Jeffrey Sachs (2008) and Nicholas Stern (2007, 2009) dominate the discussions on climate change. Sachs (2008), for example, argues that with the proper management of resources, technologies and politics we can solve the concomitant challenges of global poverty, climate change, conflict, mass migrations and financial and economic crises. His argument is optimistic, top down and favours a technocratic set of solutions. It represents an economic analysis, an economic explanation and an economic set of solutions to climate change. I suggest that we must be extremely reluctant to accept these top-down easy answers to a set of challenges that are messy and conflictive. Both adaptation and mitigation strategies require 'deliberative change and decision making about resources, values and priorities'(Adger *et al.*, 2009: xiii). A pervasive economistic and market-based perspective to climate change and poverty cannot adequately address these issues.

The most important global development agencies are rapidly building up expertise on climate change related issues and have moved work on environmental issues up in their lists of priorities. They are creating new departments and sections, and dedicating their flagship publications to the issue. The World Bank, for example, recently created a new subsection on climate change under its 'environment' section, and in the autumn of 2008 issued a Strategic Framework for the World Bank Group (WBG) on the issue of climate change. The document states that this organisation can no longer avoid addressing climate change as part of its ongoing work on poverty reduction and development, but that its mandate is to keep the focus on building growth and reducing poverty. It calls for immediate increases in aid flows. The strategy thus aims to 'enable the WBG to effectively support sustainable development and poverty reduction at the national, regional, and local levels, as additional climate risks and climate-related economic opportunities arise'

(World Bank, 2008: 5). The document places strong focus on expanding market and business opportunities for energy efficiency, but stresses that 'resources will not be diverted from financing core development needs' (World Bank 2008: 5). At the same time, the Bank had earlier created a set of Climate Investment Funds (CIFs) dedicated to supporting projects with additional funds from diverse donors. The document (World Bank, 2008) outlines six main areas of work:

1. Support climate actions in country-led development processes;
2. Mobilise additional concessional and innovative finance;
3. Facilitate the development of market-based financing mechanisms;
4. Leverage private sector resources;
5. Support accelerated development and deployment of new technologies; and
6. Step up policy research, knowledge and capacity building.

The document conveys the possibility that climate change may also be an opportunity for increased growth and economic activities; it argues for the role of markets and private sector as crucial and makes a plea for the WBG to be a key global player on the relations between climate change and development. The WBG will play a global role, we read, as 'a knowledge provider, a facilitator of North–South and South-South cooperation, a partner of global institutions, and an advocate of an efficient and just global climate policy implemented through neutral and well-governed processes and institutions' (World Bank, 2008: 6).

While this new focus on climate change by the Bank is welcome, it can also be very problematic and tricky. On the one hand, major policy prescriptions are likely to follow the strategic document just briefly presented; on the other, some departments of the Bank are putting forward more radical ideas, such as a focus on pro-poor adaptation. This is becoming increasingly the territory of the Social Development Department, which has created a new subsection on the social dimensions of climate change and produced sets of interesting and relatively progressive papers. For example, the former head of the social development department, Steen Jorgensen, co-authored a paper that argues for an important role of social protection mechanisms and for a 'no regrets' approach in addressing human vulnerability to climate change (Heltberg *et al.*, 2009). Yet although poverty and vulnerability are closely related, they are not synonymous (Eriksen and O'Brien, 2007). Poverty reduction does not automatically reduce the vulnerability of the poor. Similarly, not all types of climate-related adjustment will reduce the vulnerability of the poor; in some cases they could even increase the vulnerability of some groups (Ulsrud *et al.*, 2008).

Although the social development department of the World Bank has posed the question '*adaptation for whom?*' this is not where the core messages from the Bank emerge. Rather, core messages are most often conveyed though the flagship publication of the Bank, the World Development Report (WDR; World Bank, 2010).

The WDR in 2010, dedicated to Climate and Development, frames its main message on this issue along very similar lines to the strategy document. The slogan of the report is that 'a climate-smart world is within reach'. Developing countries can shift to lower carbon paths while promoting development and reducing poverty, a widely distributed summary claims, 'but this depends on financial and technical assistance from high-income countries' (World Bank, 2010). Although it is not the aim of this chapter to offer an analysis of the WDR 2010, it is very clear that the report fails to develop some of the more interesting and pro-poor oriented themes already visible in the social development department, treats climate change as a technical and economic issue, fails to question models of development endorsed by the Bank for decades, and even places a strong burden of responsibility for mitigation on developing countries themselves. One thing is clear, the Bank is on its way to becoming the expert institution for climate change and development, in the same way it became the expert institution for poverty issues in earlier decades.

The WDR 2010 contrasts with an earlier report presented by the United Nations Development Programme (UNDP), a pioneer institution in developing linkages between climate and development. The Human Development Report (HDR) 2007/ 2008 was focused on *Fighting Climate Change: Human Solidarity in a Divided World*. This report presents progressive ideas related to climate change and human development, and makes explicit links to human rights and human security, which are indeed consistent with the earlier work of the UNDP's Human Development Office. It builds on a development philosophy that is very different from the Bank, albeit a lot less influential in generating policy responses and actions in client countries (St.Clair, 2004; Murphy, 2006; McNeill and St.Clair, 2009). Unlike the World Bank, the HDR clearly states that the most fundamental challenge posed by climate change is the way we think about progress. It uses the threat of climate change as evidence that economic development and human development are very distinct issues and that what is at stake is the security of humanity, building on a notion of human security also pioneered by UNDP and elaborated on through Amartya Sen's capability approach, which provides the intellectual basis for human development and versions of human security. It represents a development ethic that runs counter to what I have referred to as the dominant poverty discourse (see Gasper, Chapter 2).

There could be no clearer demonstration than climate that economic wealth creation is not the same thing as human progress. Under the current energy policies, rising economic prosperity will go hand-in-hand with mounting threats to human development today and the well-being of future generations. But carbon-intensive growth is symptomatic of a deeper problem. One of the hardest lessons taught by climate change is that the economic model which drives growth and the profligate consumption in rich nations that goes with it, is ecologically unsustainable.

(UNDP, 2007/2008: 15)

What the HDR 2008 tells us is that rather than mainstreaming climate change into ongoing development projects, *we must question our assumptions about progress and development in the first place*. This is the fundamental message that is precisely missing from the much more powerful and influential actor, the World Bank Group. The most destructive effect of many poverty reduction strategies in the past and present is that they allow us to believe that poverty is a problem of aid, and thus for the aid community to solve. As Victoria Lawson and I have argued (Lawson and St.Clair, 2009), poverty studies have mainly focused on the global South, and have been dominated by economistic views based on neoliberal premises that promote market-based solutions. It has succeeded in presenting expertise on poverty issues only when the issues are seen as an economic matter, analysed and researched taking as departure points narrow quantitative and aggregative definitions of poverty. Dominant poverty knowledge is also methodologically individualistic, constructing 'poor people' as a category that stands outside of social relations. Such perspectives disregard history, which reveals the role of economic and political power differentials in creating and perpetuating poverty; the agency of people and collectivities and the role of culture or religion; and the ethical and social justice aspects of poverty. It also privatises the possible policy solutions and does so very often by situating the task of poverty reduction within the goals of global capitalism. One of the most pervasive and perverse consequences of this economistic and neoliberal view of the poor is that it ignores processes of accumulation and wealth creation that have produced and perpetuated situations of poverty across the globe.

Thus, a proper understanding of the relations between climate change, poverty and development calls, first and foremost, for a critical investigation of the failures, limitations and problems of mainstream and dominant poverty and development knowledge. The fact is that most inhabitants of this planet are severely poor, chronically poor or vulnerable to poverty; poverty is increasing rapidly in advanced economies, middle classes are shrinking in most parts of the world; and social protection mechanisms have been dismantled, privatised, made inaccessible to poor and vulnerable people. Aid contributions are outrageously low, slow to produce results and based on misguided development paradigms. The twin challenges of reducing poverty and responding to climate change demand very different types of actions, including a deep and transformative reflection on the meaning of development. Mainstreaming climate concerns into development aid work will provide neither results nor the needed support from countries and citizens of the global South.

If the relationships between climate change and poverty reduction are framed along the lines seen thus far (as a question of development and adaptation aid), we will miss the extent to which these issues are symptoms of misguided ideas of progress and modernisation. As Wendy Harcourt puts it:

What is needed is a systemic change. Climate change forces us to see that the dominant development model is not working. It is not enough for individuals in the North to start recycling, to buy 'eco' and 'bio' products, to pay a little more on the air ticket to 'offset' costs, to turn off the taps and lights to 'save' water and energy. None of this will be enough. The emphasis on individual behaviour only distracts attention away from corporations and governments, who determine policies on a daily basis that lead to massive scale ecological destruction.

(Harcourt, 2008: 307)

We will also miss the extent to which these two issues are both symptoms of a 'wider social malaise'. Verweij and Thompson put it bluntly:

The way humans pollute, degrade and destroy the natural world is merely a very visible indicator for the way they treat each other and particularly weaker members of society. The logic that allows us to fell thousands of square kilometres of rainforests, to dump toxins in waterways or pollute the air is precisely the same logic that produces racism, misogyny and xenophobia. Tackling one problem inevitably implies tackling all the others.

(Verweij and Thompson, 2006: 9)

Similar views are expressed by contributors to this volume. Given the status of the scientific knowledge that we already possess, and if the global community is serious about adaptation without neglecting the need to mitigate climate change, then the most efficient way to build adaptive capacity may be to immediately eradicate severe poverty and to focus our political, economic and research attention and capacities on thinking about and drafting feasible ways to meet what in fact is a very basic Western moral idea: that we should protect people from known harms. To continue the business of poverty reduction as it has been carried out over the past decades seems indeed to be a road to nowhere.

The importance of global knowledge institutions

Whose knowledge counts and which voices and framings are heard in the relations between climate change and poverty is a fundamental question. Part of the task is to understand more deeply the role of the knowledge producers that specialised global institutions have and to evaluate the extent to which their views are the result, or not, of democratic processes of tightly monitored peer review processes and high academic standards. The World Bank, UNDP and the IPCC are similar in regards to their role as knowledge producers; but they are very dissimilar in regards to the processes of building such knowledge and the legitimacy and credibility it may have among diverse stakeholders. When the World Meteorological Organization (WMO) (established in 1950) and the United Nations Environmental Programme (UNEP) (established in 1972) jointly created the Intergovernmental Panel on Climate Change (IPCC) in 1988, their goal was to create a knowledge institution, but one that does not carry out research. IPCC's goal is to 'assess on a comprehensive, objective, open and transparent basis the scientific, technical and socioeconomic information relevant to understanding the scientific basis of risk of human-induced climate change, its potential impacts and options for adaptation and mitigation'. The IPCC does not carry out new research, nor does it monitor climate data or other relevant parameters. It bases its assessment mainly on peer reviewed and published scientific/technical literature.

The reports of the IPCC have served as a basis for global environmental policy. The Kyoto Protocol, for example, was based on the first two IPCC assessments. It has taken two more assessments (and the help of a Nobel Peace Prize), to counter-balance the accusation of poor science that the IPCC reports have had in the past two decades. It is still the belief of many that bad weather, such as hurricanes, droughts, floods and other 'natural disasters', and their impact on the poor and vulnerable are the result of God's hand or a temporary small natural variation in the world's climate. Not only lay people, but highly prominent politicians and public officials in positions of responsibility have dismissed the IPCC as a producer of bad science. Yet this disbelief is at odds with the institutional structures in place for producing climate change knowledge. The IPCC was considered by a report of the US National Academies (2002), *Knowledge and Diplomacy: Science Advice in the United Nations System*, to be the best well-functioning UN institution producing science for policy. The IPCC has the most sophisticated peer review mechanisms of any of the UN agencies or even any domestic public institution.

Without suggesting that the IPCC has no problems related to dominant views and discourses, it is blatantly clear that the institutions that produce the most widely accepted knowledge on poverty, the UN System institutions, donors and the World Bank, have achieved nothing compared to the triple global peer review mechanisms that underlie the IPCC assessments. The role of expert knowledge in global problems that are ill-structured, global, historically rooted, difficult to analyse in terms of causality and with diffused relations of responsibility and highly politicised because of vested interests such as climate and poverty has been discussed in earlier work (St.Clair 2006a, 2006b; McNeill and St.Clair 2009). This work suggests a clear dominance of World Bank perspectives in the field of poverty due to many factors unrelated to the actual quality and democratic value of such knowledge. Even if many of the UN System organisations and scholars across the globe produce and dissem-inate important knowledge of different nature and with diverse policy recommenda-tions to the Bank, they do not have sufficient force to counterweight the Bank's perspectives. Alternative perspectives and voices, such as those from activist com-munities or critical scholarship, are often dismissed as non-knowledge or irrelevant.

Thus, unlike the IPCC in relation to climate change, there is no coherent body of knowledge resulting from triple peer review processes of published work worldwide on matters related to poverty and development. And in the absence of a global knowledge institution assessing the state-of-the-art understandings of poverty – an institution analogous to the IPCC, expert institutions like the World Bank are enable to both frame and drive issues in particular directions.

Clearly, the IPCC itself is poorly prepared to now include social science based knowledge on poverty issues in its assessments. Perhaps a factor contributing to this failure is precisely the lack of a parallel body for poverty and development research, with which the IPCC could have engaged in dialogue and collaborative work. But I suggest that in the absence of such dialogue, the IPCC has adopted the language of what is de facto an incoherent expert community, and primarily the ideas of the dominant expert institution (such as the World Bank) that has less rigorous standards in the ways in which it formulates and distributes knowledge, and thus lacks the credibility that the IPCC demands in its own field. The end result is an already dominant view that presents a narrative depicting climate change in its relation to poverty as a technical and environmental issue dislocated from the human aspects, social relations and social institutions, ludicrously value-free, even if often wrapped up in ethical rhetoric.

A good way to envision what happens when environmental change is dislocated from social relations and when ideas of progress and development and their relations to poor are transformed into policy tools is to learn from history. I suggest we listen carefully to historian Mike Davis' powerful account of relationships between global environmental change, progress and poverty during the Victorian era (Davis, 2002). In *Late Victorian Holocausts: El Niño Famines and the Making of the Third World*, Davis presents a political ecology of famine that documents and illustrates that each global drought was the green light for an imperialist landrush (Davis, 2002). He shows that moral, political–economic and knowledge variables played against each other, resulting in death and impoverishment that is still felt today. Extreme poverty was thought to be caused by bad weather, which disrupted agricultural practices and thus affected the crops that provided the survival mechanisms of the most vulnerable sectors of society. The thesis that famines were caused by weather patterns permitted Victorian viceroys, bureaucrats and colonists as well as the British Empire's domestic economic and political forces to distance themselves and their actions in their colonies, from the famines and persistent poverty that they were witnessing. When bad weather intensified, the resulting famines had, in the eyes and minds of Victorian viceroys and their peers back in Britain, a clear causal explanation. The thesis that severe poverty and famines were caused by climate freed their consciousness and erased/obscured their accountability towards their constituencies from any moral responsibility; countries of the South, affected by disadvantaged weather conditions, were lands of poverty and famines. These 'disadvantaged' weather conditions were caused by God, rather than the prevailing social relations, political economy or outright colonialism.

One of the fundamental lessons of Davis' analysis is that the disassociation between nature and society, and between environmental change and poverty, follows a trajectory that is both historical and political. Such disassociation relates to the dominance of the interests and misuse of the dominant morality of powerful groups; it is related to the development of capitalism, modernisation and industrialisation; and it is related to the takeover of economics as the dominant science for explaining and predicting change. These were a heuristic technique that served the interests of the powerful, to the disadvantage of the vulnerable. Indeed, there is a well-established tendency to reject, obscure or ignore data and theories that show that advanced economies and dominant groups are harming others. This includes a historical pattern in the rejection of scientific knowledge showing economic policies and development strategies of advanced economies have causal relationships to the life chances and opportunities of poor groups and poor regions, and a historical pattern in the rejection of scientific knowledge that challenges the lifestyles and ways of living that maintain and perpetuate the dominance of advanced economies and privileged groups. Ethics and morality concerns are entangled with such rejections of some scientific knowledge over other alternatives. There is a well-established tendency to separate 'us' (advanced economies or well-off groups) from 'them' or 'the others' (those in poor countries or poorer sectors of our own countries), and to presume instead that all is well with 'us' and shift personal blame on those groups that are deprived and marginalised. This pattern continues today, primarily driven by processes of knowledge production that are far from democratic, integrative and fair. In spite of progress in the processes of building knowledge on climate change, there has been no parallel progress in the field of poverty.

Often, the same people and groups who look at the results of the IPCC's Fourth Assessment Report with scepticism, do not stop to question the, at least equally, problematic scientific approaches, modelling and predictions regarding the causes and the solutions to global poverty. Any scientific approach that challenges the view that more trade, more neoliberal economic policies, more privatisation of social services, etc. may not be good for the poor is challenged as bad science. Social democratic-like state intervention, policies promoting egalitarian societies (equalising income, high taxation and universal social policy, to name just some anti-poverty measures that have been successful in the reduction of poverty in Norway) are dismissed as irrelevant knowledge. Yet the dominant economic science that drives poverty knowledge is at least as worthy of debate as the modelling and predictions of current climate change science.

A small but significant example is related to food security. For decades the world has witnessed the destruction of small-scale farming and the take over of farming by corporate-led and market-based agriculture. Yet poverty experts continue to defend the suitability of corporate agriculture, even in the face of major environmental changes and food shortages. Meanwhile, alternatives are dismissed, despite their success and potential being well documented. Debates on food security continue to be dichotomised between those supporting corporate agriculture as solutions to hunger versus those calling for radical dismantling of this system. While the former dominates, success stories from alternative food systems continue to be cited. There are success stories on the rich experiences of small farmers, peasant communities and cooperative systems that have achieved food security along with equity and ecological sustainability. As Holt-Giménez *et al.* (2009) argue, groups like Via Campesina, The Landless Movement (MST) in Brazil and small-scale agriculture in the majority world present lessons of the inseparability of economic organisation, technology, equity, sustainability and democracy (Holt-Giménez *et al.*, 2009).

There are important lessons for further knowledge about the relations between climate change and poverty as well as for the types of ethical arguments that could be made and those likely to succeed in shifting the discourse. Politics and the perpetuation of domination and power of certain global knowledge actors have much to do with the science versus non-science debate as we are witnessing today on the poverty front. We could have a history of scientific knowledge about the environment and poverty that is more aware of the causes and consequences of one another. We could have a science that prevented rather than intensified poverty, and that produced solutions for mitigation and adaptation a long time ago (e.g. from funding and promoting alternative energy research decades ago, to creating warning systems and disaster prevention for vulnerable areas, to creating social protection mechanisms intended to generate egalitarian societies, preventing both excessive accumulation of wealth and deprivation). My point is that now there is an opportunity to build alternative knowledge that is much more aware of these relations. We must learn from the past and current failures to define, measure, address and prevent poverty in both the North and the South, and from the lessons of disassociating poverty from the environment. We must also learn from the strategies that have permitted a dominant knowledge to prevail, a particular type of discipline to dominate the definition of what is and what is not scientific knowledge on poverty and who are the poor, and to challenge those strategies. There is also an opportunity to learn about the failures of ethical thinking to challenge such actions, and about the ways in which distorted accounts of the facts become a way to justify the morality of the rich, to blind us towards the suffering of others, and how ethical rhetoric added to such distortions is often a reason for preventing rather than encouraging serious debate about responsibilities.

Towards the responsibility to protect people from poverty and climate change

It is possible to envision a near future where the negative impacts of climate change on poor societies, communities and people becomes commonplace, viewed on television screens and Internet blogs by those who are more fortunate. In the same way as we have become used to the spectacle of poverty and the suffering of distant others, we are unable to see the immorality of being unresponsive even if able to respond (Sontag, 2004). Will we become unresponsive spectators of climate change suffering – people dying of hunger due to food shortages, dying or suffering from floods, refugees and internally displaced people unable to claim protection from other countries, in the same way that we are spectators of the death and suffering of millions of people because of poverty conditions or refugees and displaced people for other reasons? If the carrying capacity of the Earth is to be reduced to no more than 1 billion people by the end of the twenty-first century, as James Lovelock envisions, then we should consider who will be the ones to survive?

The complexity and scope of these problems is so enormous that human beings tend to disconnect from their own relations and personal links to these problems. This disconnect is reinforced by the way in which dominant knowledge transforms the issues into technical matters to be left to experts, in this case the expert bodies of development aid. It is relatively easy for well-off citizens of advanced economies or for elites and well-off members of poorer countries to continue a system that demands very little from them. The disconnect is widened by the underdevelopment of political philosophy, responsibility theory and the shortcomings of a methodologically territori-alist Western ethical theory without a solid tradition for dealing with problems of global scope (see also Gardiner, Chapter 8). Some may see a focus on responsibility as misguided precisely because of this complexity and the lack of theoretical tools.

But the human imagination is endless. The global community has solved many issues when it is able to overcome individualism and selfishness. Global institutions are in place that could be reformed, changed or transformed. Framings and dis-courses on poverty and development that have already incorporated normative foundations, such as human development, human security and human rights based views, could serve as a point of departure for further efforts to implement alternative measurements, indicators and the needed global institutions that can materialise their principles. There is much to be learned from the field of development ethics, as this interdisciplinary field of research has challenged unsustainable and inequitable visions of development for decades. The capabilities approach represents one alternative view on development, considering it as an ethical challenge with ethic-ally grounded analysis and policy recommendations. What the field of develop-ment ethics has learned is that when conflict and broad-based debate is forged, alternatives flourish (St.Clair, 2007; Crocker, 2008; Gasper and St.Clair, 2010). Social justice and development that enables people to live the lives they have reasons to value, as Sen rightly argues, requires social arrangements, but also the creation of spaces for value disagreement and reasoning (Sen, 1998, 2009). Development ethics offers us many years of struggling with difficult questions.

For example, the understanding that one of the most important ways to control demographic changes is via gender equality, women's empowerment and universal access to high quality education is a result of such views. And many of the principles advocated by ethicists have found place in alliances with progressive global institutions. A strong focus on universal access to social protection mechanisms, such as the one proposed by a coalition of progressive UN institutions led by the International Labour Organization, has much to offer for a rethinking of poverty research and policy. The Global Social Floor proposal that is now emerging can be a point of departure for rethinking how to minimise the negative impacts of climate change in vulnerable populations while serving to also protect people from the worst consequences of poverty itself.[2]

We have the possibility to create spaces of action at the local level, to tap into people's innovation and capacity to cope and to be resilient. There are many successes on the ground, particularly in cases where the collective actions of citizens have been at the centre of poverty reduction work and where solidarity and responsibility has framed the issues (Green, 2008). Disparities of interests and understandings of same issues should not lead to despair and cynicism. Some advocate for democratic deliberation and for bottom-up democracy (Korten, 2009); others for clumsy solutions (Verweij and Thompson, 2006). And we have accounts of experiences where inelegant, yet effective, solutions to complex problems on the interface between environment and humans have succeeded (Holt-Giménez *et al.*, 2009).

Alternative perspectives can be focused, as earlier sections have suggested, on lessons from history on the relations between climate variability and poverty conditions, political economy, that could prevent rather than promote famines and billions of preventable deaths. We can envision research that centres it attention of the factors that contribute to build social cohesion and egalitarian societies and that take as objects of study not the poor but rather elites and the middle classes. Such perspectives can then be a means to forge a sense of 'us', rather than reinforcing the 'I' so common among well-off citizens of liberal democracies. A focus on immediate collective action for abolishing severe poverty and protecting people from falling into it would have to tap on all those ideas and many others. Evidently, most of all these transformations in perspective and transformational ideas would take time if they are left to follow the usual channels for building the needed legitimacy and credibility, and if they are to achieve the needed policy impact.

---

[2] The case for a basic social floor is outlined in a briefing paper 'Can low-income countries afford basic social security?' Available online: http://www.ilo.org/public/english/protection/secsoc/downloads/policy/policy3e.pdf. See also 'GSP Digest.' *Journal of Global Social Policy*, 2008: 8(3); and Michael Cichon (2008) 'Building the Case for a Global Social Floor.' Available online: http://www.un.org/esa/socdev/social/documents/side%20events/ILO_Building_the_case.ppt.

Thus, we must understand these as fundamental instruments to lead towards a global deal to stop catastrophic change. A global deal is only viable with the joint work and collaborations of the majority world. The West can no longer drive the agendas, nor drive the processes of building the only legitimate knowledge.

The focus on the immediate abolition of poverty has value as a process in building the much-needed solidarities within and across groups and countries that long-term mitigation and adaptation require. The 'other' (the poor in far away lands or the unborn future generations), becomes part of the 'us' (all of us inhabitants of Gaia and members of the human race). To forge this 'us' requires consensus building across social groups and active public debate to confront what are undeniable conflicts of interests and perspectives. As the time is short, we have a decade at the most to make the radical changes required to avoid irreversible changes in the Earth System, an immediate focus on responsibilities is needed and the responsibility to protect is the way to narrow down the seeking of feasible answers.

Pro-actively seeking ways to frame forward-looking and feasible forms of political responsibility will guide us in finding ways to fairly balance rights, duties, obligations and uncertainties. The focus on responsibilities is a much better point of departure for agreements and negotiations at the global level; it jointly addresses climate and poverty, it gives a direction, if not a road map, complementing the directionality of the concept of human security and helping to materialise its principles. The materialisation and outlining of specific duties and responsibilities will have to occur at many levels, but it frames the issues as a problem for all human beings and is about all of us. It is not an 'elegant' solution or a managerial perspective. Allocations of responsibilities are always messy matters, leading to constant processes of negotiations and debate as disagreements about responsibility are very often disagreements about interpretations of facts. It is a 'clumsy' yet a directional force. What matters, I argue, is keeping the debate about responsibilities open-ended, ongoing, and again, as a direction for action. A good way to start discussing responsibility is with those actors, agents and individuals with the most power and thus the most ability to respond to concrete issues (response-ability). A major reform of the global institutions is the first point in this responsibility agenda. The twenty-first century can only 'be' if it becomes the 'age of responsibility'.

References

Adger, W. N., Lorenzoni, I. and O'Brien, K. L. 2009. *Adapting to Climate Change: Thresholds, Values, Governance*. Cambridge, UK: Cambridge University Press.

Bond, P., Dada, R. and Erion, R. 2008. *Climate Change, Carbon Trading and Civil Society Negative Returns on South African Investments*. Scottsville, South Africa: University of KwaZulu-Natal Press.

Crocker, D. A. (2008). *Ethics of Global Development: Agency, Capability, and Deliberative Democracy*. Cambridge, UK: Cambridge University Press.

Davis, M. 2002. *Late Victorian Holocaust: El Niño Famines and the Making of the Third World*. London: Verso Publishing.

Deacon, B. 2007. *Global Social Policy and Governance*, London: Sage.

DFID (Department For International Development) 2009. *White Paper: Building our Common Future*. London: Department For International Development.

Eriksen, S. and O'Brien, K. 2007. Vulnerability, poverty and the need for sustainable adaptation measures. *Climate Policy*, **7**, 337–52.

Gasper, D. and St.Clair, A. L. 2010. The field of development ethics: Introduction. In D. Gasper and A. L. St.Clair, eds., *Development Ethics: A Reader*. London: Ashgate.

Gough, I. *et al*. 2008. JESP symposium: climate change and social policy. *Journal of European Social Policy*, **18**, 325–44.

Green, D. 2008. *From Poverty to Power: How Active Citizens and Effective States can Change the World*. Oxford, UK: Oxfam International.

Harcourt, W. 2008 Editorial. Walk the talk: putting climate justice into action. *Development*, **51**(3), 307–309.

Heltberg, R., Siegel, P. and Jorgensen, S. 2009. Addressing human vulnerability to climate change: toward a 'no-regrets' approach. *Global Environmental Change*, **19**, 89–99.

Holt-Giménez, E., Patel, R. and Shattuck, A. 2009. *Food rebellions! Crisis and the Hunger for Justice*. Cape Town: Pambazuka Press.

Hulme, M. 2009. *Why we Disagree About Climate Change: Understanding Controversy, Inaction and Opportunity*. Cambridge, UK: Cambridge University Press.

Korten, D. C. 2009. *Agenda for a New Economy: From Phantom Wealth to Real Wealth*. San Francisco, CA: Berrett-Koehler.

Lawson, V. and St Clair, A. L. 2009. *Poverty and Global Environmental Change. International Human Dimensions Programme Update*. Issue 2. Available online: http://www.ihdp.unu.edu/category/47?menu=61.

McNeill, D. and St.Clair, A. L. 2009. *Global Poverty, Ethics and Human Rights: The Role of Multilateral Organizations*. London and New York: Routledge.

Murphy, C. 2006. *The United Nations Development Programme: A Better Way?* Cambridge, UK: Cambridge University Press.

NOU 2009. *White Paper. Climate, conflict and capital: Norwegian development policy in a new environment*. Norwegian Ministry of Foreign Affairs Report No. 13 (2008–2009). Available online: http://www.regjeringen.no/en/dep/ud/Documents/Propositions-and-reports/Reports-to-the-Storting/2008–2009/report-no-13–2008–2009-to-the-storting.html?id=552810.

OECD 2002. *Poverty and Climate Change: Reducing the Vulnerability of the Poor Through Adaption*. Available online: http://www.oecd.org/dataoecd/60/27/2502872.pdf.

OECD 2008. *Statement of Progress on Integrating Climate Change Adaptation into Development Co-operation*. Available online: http://www.oecd.org/dataoecd/44/53/40909509.pdf.

Sachs, J. 2008. *Common Wealth: Economics for a Crowded Planet*. New York, NY: Penguin Press.

St.Clair, A. N. 2004 The role of ideas in the United Nations Development Programme. In M. Bøas and D. McNeill, eds., *Global Institutions and Development: Framing the World*, London: Routledge.

St.Clair, A. L. 2006a. Global poverty: the co-production of knowledge and politics. *Global Social Policy*, **6**(1), 57–77.

St.Clair, A. L. 2006b. The World Bank as a transnational expertised institution. *Global Governance*, **12**(1), 77–95.

St.Clair, A. L. 2007. A methodologically pragmatist approach to development ethics. *Journal of Global Ethics*, **3**(2), 143–64.

Sen, A. K. 1998. *Development as Freedom*. New York, NY: Knoop.

Sen, A. K. 2009. *The Idea of Justice*. New York, NY and Cambridge, UK: Belknap Press/ Harvard University Press.

Sontag, S. 2004. *Regarding the Pain of Others*. New York: Picador.

Stern, N. H. 2007. *The Economics of Climate Change: The Stern Review*. Cambridge, UK: Cambridge University Press.

Stern, N. H. 2009. *The Global Deal: Climate Change and the Creation of a New Era of Progress and Prosperity*. New York, NY: Public Affairs.

Ulsrud, K., Sygna, L. and O'Brien, K. 2008. More than Rain: Identifying Sustainable Pathways for Climate Adaptation and Poverty Reduction. Oslo: The Development Fund. Available online: http://www.utviklingsfondet.no/Utviklingsfondet_-_forsiden/ Temaer_vi_jobber_med/Klimaendringer_i_Sor/More_than_rain/.

UNDP (United Nations Development Programme) (2007). Human Development Report 2007/2008 *Fighting climate change: Human solidarity in a divided world*. Available online at: http://hdr.undp.org/en/reports/global/hdr2007–2008/

US National Academies 2002. *Knowledge and Diplomacy Science Advice in the United Nations System*. Washington, DC: National Academies Press.

Verweij, M. and Thompson, M. 2006. *Clumsy Solutions for a Complex World: Governance, Politics and Plural Perceptions*. Basingstoke, UK: Palgrave Macmillan.

World Bank 2008. *Development and climate change: a strategic framework for the World Bank Group*. Available online: http://beta.worldbank.org/overview/strategic-framework-development-and-climate-change.

World Bank 2010. *World Development Report: Development in a Changing Climate*. Available online : http://econ.worldbank.org/WBSITE/EXTERNAL/EXTDEC/ EXTRESEARCH/EXTWDRS/0,, contentMDK:20227703~pagePK:478093~piPK:477627~theSitePK:477624,00.html.

11

Social contracts in a changing climate: security of what and for whom?

BRONWYN HAYWARD AND KAREN O'BRIEN

Introduction

Climate change is exacerbating complex social–ecological changes that threaten the security of individuals and local communities around the world. The complexity and the sheer scale of the risks now associated with climate change has undermined the idea that any nation state, acting alone, can credibly claim to provide security for its citizens (Pelling and Dill, 2006, 2009; O'Brien, Hayward and Berkes, 2009). The global extent of climate change, with its diverse local manifestations, calls for rapid and comprehensive responses at all levels to reduce both greenhouse gas emissions *and* vulnerability to climate variability and change.

Given these challenges, is not surprising that there are urgent calls for new political solutions. A new 'social contract' is proposed by many as a way to organise a more effective, collective response to climate change. Individual politicians and non-governmental organisations alike have called for new agreements between citizens and the state to address the threats associated with climate change. For example, in 2006 David Miliband (then Environment Secretary in the UK), argued that a new 'environmental contract' is essential to clarifying the rights and responsibilities of citizens, businesses and nations to one another (Miliband, 2006). In 2008, members of the European Parliament called for a new 'global contract' for climate justice, to promote environmental effectiveness, avoid unduly harsh economic impacts and 'shield the world's poor' from the worst effects of climate change (Edenhofer *et al.*, 2008). Non-profit think tanks, such as the New Economics Foundation, have argued for a 'green new deal' to tackle the 'triple crunch' created by the financial crisis, climate change and peak oil (Green New Deal Group, 2009). The latter organisation has called for a rethinking of contract ideas, capturing the public mood for 'a modern translation of the politics of hope and pragmatism employed by Roosevelt in the 1930s arguing that [t]hen, as now, someone need[s] to pick up the pieces of a system failed by short-termism and unenlightened self-interest' (Simms, 2008).

Climate Change, Ethics and Human Security, eds. Karen O'Brien, Asunción Lera St.Clair and Berit Kristoffersen. Published by Cambridge University Press. © Cambridge University Press 2010.

Contemporary contracts to address climate change are being discussed and developed in an atmosphere of economic crisis and environmental fear, and within a context of dynamic and diffuse global power relationships. These new contracts are rarely exposed to sufficient, inclusive public debate, and are therefore unlikely to secure equity or justice for marginalised groups or non-human nature. There is a need for greater reflexivity in relation to emerging responses to climate change, particularly in relation to two key questions: What is it that is being secured? And for whom?

Social contract theory, which involves a tacit or explicit compact or agreement about reciprocal rights and duties between citizens and authorities, has been highly influential in liberal democratic thought (Weale, 2004). However, one could also argue that social contract theory lies at the heart of the current environmental crisis, serving as a legitimising tool for power relationships that have perpetuated human injustice, unsustainable resource extraction, colonisation and state expansion (see Dobson, 2003; Eckersley, 2004; Weale, 2004; Nussbaum, 2006; Pateman and Mills, 2007). The effect of contractual arrangements has often been to subordinate and exploit nature and the physical environment in the name of development, progress and economic growth (Kissi-Mensah, 2008).

In this chapter, we consider emerging contracts as a response to climate change from a critical perspective. We argue that the 'centred' nature of these new contracts, whereby power and decision-making capacity remain concentrated within a clearly defined nation state, or group of nations, is likely to promote unsustainable growth and exacerbate inequality by accelerating the private appropriation of ecosystem services. Indeed, such an approach tends to displace and obfuscate the justice issues that are at the very heart of anthropogenic climate change (Adger *et al.*, 2006). Below, we first briefly describe social contract theory and critically examine its role in shaping liberal expectations of governance in a changing climate. We then review contemporary calls for new 'environmental' or 'climate' contracts and consider whether 'tweaked' social contracts will provide greater human security in a warming world. Finally, we identify alternative visions for a transformative eco-social compact, inspired by the writings of eighteenth-century social contract theorist, Jean-Jacques Rousseau, and, more recently, by Sen and Nussbaum's capability approaches (Nussbaum and Sen, 1993; Sen, 1999, 2009; Nussbaum, 2006), Dobson's model of ecological citizenship (2003) and Eckersley's green state (2004). These alternative visions resonate with contemporary calls for a more explicit discussion of morality in environmental decision making, less emphasis on securing economic growth, greater emphasis on securing human capacity to flourish within ecological limits and more procedurally just decision making (Adger and Jordan, 2009; Jackson, 2009; Sippel, 2009). We argue that an alternative social contract or eco-social compacts contains the seeds for a more equitable and sustainable way to foster human security in a changing climate.

Social contract theory: a problematic legacy for sustainability

The concept of a social contract has been highly influential in shaping Western thinking about what makes government legitimate, and the respective roles and responsibilities of government and citizens. From Aristotle to Rawls, political thinkers have wrestled with ways to develop agreements that legitimate governments, provide for social order, protect human well-being and foster the conditions for human flourishing (Kant, 1959; Locke, 1965; Rousseau, 1966, 1968, 1973; Rawls, 1971; Hobbes, 1998). Despite diverse visions, social contract theorists generally agree that legitimate, collective governance arrangements are informed by the consent of the people (Harsanyi, 1976; Gauthier, 1986; Barry, 1995; Scanlon, 1998; Weale, 2004; Pateman and Mills, 2007). The notion of government by consent is simple and powerful and helps explain why the ideas of social contract theory have continued to have an important influence on modern governments (Pateman and Mills, 2007). For example, based on the idea of a social contract, a state may be obligated to provide order and security, such as upholding civil and political freedoms or protecting the right to own property, in return for a citizen's agreement to support government (e.g. paying taxes, engaging in civil and law-abiding behaviour and voting).

Given the complexity of the challenges associated with climate change, however, the idea that large-scale democratic systems can hope to 'solve' the problem using the ideas of a social contract framework that evolved over centuries may seem at best a quaint, optimistic hope; at worst, a misleading, obfuscating ambition that disguises the very power inequities it purports to control and enables the undermining of socio-ecological systems to continue unchecked (O'Brien *et al*., 2009). Nonetheless, the concept of the social contract has flourished, not withered, as a tool for governance in a changing climate. As Boucher and Kelly (1994) note, the ideal behind social contracts (i.e. of an agreement on fair terms of association between individuals who share a recognised, free and equal status) resonates deeply with modern culture and has proven to be a great inspiration to those who do not enjoy the recognition of that status. Yet as the reach and grasp of market-liberal contracts extends to wider and wider groups, including to distant populations whose resources have been unjustly appropriated, and to future generations whose climate is being irreversibly altered by current human activities, there is a renewed need to deliberate and contest new contracts emerging in response to climate change.

While social contract theory has inspired many, the contractual relationship itself can be considered problematic. Since the seventeenth century, social contract thinking has been allied with a liberal world view, based on the belief that private individuals are motivated to engage in political action by self-interest, which can be regulated through market-like contracts (Macpherson, 1973). Shaped by the

Enlightenment project, and responding to the aspirations of a newly emerging property-owning class, social contract theory reflects a vision of citizenship based on utility maximisation and consumption (Macpherson, 1973), embedded values about the desirability of economic growth, and a defence of the rights and freedoms of property owners from undue government intervention (Weale, 2004). Viewed in this light, the evolution of social contract thinking has privileged economic growth and protected the rights of a minority of property owners and investors at the expense of the wider community and non-human nature. Critics also argue that this 'contractarian' perspective constrains the political imagination about other possible relationships or ways of interacting in public life because it is based on a limited view of what motivates interactions, which emphasises market liberal values that at best ignore the interdependent relationship between humans and the natural world, and at worst actively condone human exploitation of natural resources (M. Midgley, 2003; D. Midgley, 2005). The natural world has not been regarded as party to the social contract, but as a source of resources to meet human needs (Latta and Garside, 2005).

Social contract theory has inspired some of history's most successful attempts to extend human rights to oppressed peoples. However, in practice and theory many social 'agreements' have been reached by excluding or silencing dissenting voices. Indeed, the oppression and exclusion of women and indigenous communities has been argued to be essential to the way social contracts have protected the rights and prosperity of white, male property owners through history (see Pateman 1988; Mills 1997; Pateman and Mills, 2007). Just as the evolution of the social contract has disempowered some groups, other new parties to these contracts have emerged and gained strength over time. In recent years, business organisations, in particular, have taken on new roles and responsibilities, including responsibilities formally assigned to the state, such as provision of education, water services and health care. These private businesses are increasingly global in both scale and impact, but are not accountable to the communities affected by their decisions in the way that citizens have traditionally (if inadequately) been able to hold governments to account through elections (Cragg, 2000; Eckersley, 2004; Zadek, 2006; White, 2007).

Yet business is not the only new partner in contemporary social contracts. Numerous civil society organisations have also taken responsibility for addressing critical global problems (White, 2007). Whereas the new role for businesses in the social contract was invited and encouraged by the state in a neoliberal era (Dobson, 2003), civil society has not always been a welcome partner, often being regarded more as a gate-crashing guest whose entry into new contract arrangements was facilitated by struggle and protest, rather than by invitation (Walzer, 1998). Non-governmental groups have campaigned, for example, to have wider human development priorities addressed in international agreements, arguing that climate change is also fundamentally a question of poverty alleviation (Duxfield, 2007). In light of the problematic evolution of social contract theory, discussion now turns to consider recent calls for new contractual approaches to climate change.

The 'tweaking' of the social contract in response to climate change

Global environmental problems underscore a 'conceptual crisis' now facing modern democracies (Midgley, 2005). Andrew Hurrell (2006) considers the contemporary ecological challenge as both important and profound because it calls into question the practical viability and the moral adequacy of a pluralist conception of a state-based global order. Yet, instead of challenging the political relationships that have contributed to the problem, social contract theory is increasingly being used to justify more market-based responses to global problems. Many of the contracts proposed as responses to climate change emphasise strategies such as development of and investment in low carbon technologies (Edenhofer *et al.*, 2008; Green New Deal Group, 2009). The establishment of a global carbon market, in particular, highlights the trend toward market-based solutions for climate change. Carbon agreements are less explicitly *social* (defined here as an agreement between government and people) and more explicitly *market* contracts (or agreement between individuals and businesses in the marketplace). Moreover, responsibility for implementing these market-based strategies is often directed towards established global financial institutions. For example, in the run-up to the 2009 Copenhagen talks, the executive director of the United Nations Framework Convention on Climate Change (UNFCCC) mooted the idea that the World Bank may be an appropriate institution to implement any emerging global climate agreement (De Boer, 2009).

One of the dangers of many of the proposed contracts is the seductive nature of their simple, clear rhetoric (Bohman, 2004; Eckersley, 2004; Young, 2006; Hayward, 2008). Although their elegance is persuasive, such clearly articulated solutions are often unable to secure what they promise within a complex reality (Freeden, 2009). For example, a defined group of citizens or nations may identify the emissions of carbon from the burning of fossil fuel as 'the problem' and hence agree to invest in clean technology as compensation for past and present emissions of greenhouse gases. Likewise, a group of stakeholders may agree to establish a framework for buying, selling or trading carbon emissions amongst nations as compensation for continued greenhouse gas emissions. Such agreements, however, limit the problems and often obscure complex issues, focusing only on the symptoms, rather than on the messy and complex underlying processes that contribute to climate change. In fact, such 'centred' methods of decision making may prove wholly inadequate when faced with diffuse power relationships and environmental challenges that cross national boundaries and threaten future generations (Young, 2006; Hayward, 2008; Freeden, 2009).

The proposed contract arrangements offer only minor adjustments to the three main pillars of the Kyoto Protocol (i.e. the clean development mechanism, joint implementation and emissions trading schemes). The liberal market values that underpin many of the offered 'solutions' appear to be accelerating the private appropriation of ecosystem services, often with insufficient public debate. One result is that potential 'solutions' to climate change may create new vulnerabilities. For example, an expansion of biofuel plantations may result in landgrabbing, and forestry projects for carbon sequestration may conflict with the land needs of local communities (Gundimeda, 2004). Indeed, even contracts that protect new investments in green technology may simply continue to privilege fundamentally unsustainable economic growth (Jackson, 2009). From this perspective, investments in technology may not be a solution to climate change because growth processes themselves have driven ecological degradation. The 'tweaked' environmental contracts can be charged with failing to address the very values that privilege inequitable resource consumption (Dobson and Eckersley, 2006; Jackson, 2009).

Furthermore, the track record of market contracts does not suggest that they can be easily modified to accommodate multiple scales of responses, from nation states to international and local agreements, let alone extend provisions to protect previously silenced or excluded voices. To be implemented effectively, these contracts require clearly identified parties with clearly defined rights and obligations. However, it is difficult to hold parties clearly to account for their actions. For example, in the case of carbon markets it has proved very difficult to impose an effective ceiling or cap on the emissions of parties to the contract. These problems are not limited to carbon trading, and it is not obvious how many of the new contractual solutions would be applied in the context of a changing climate, where the decisions and actions of one group of citizens can have far-reaching, detrimental impacts on distant and unknown others (Dobson, 2003; Bohman, 2004). Consequently, on closer inspection, market contracts fall far short of a transformative vision for a new and just climate regime.

In the absence of explicit consideration of the normative visions embedded in any new political arrangements, not to mention power inequities, such tweaked contracts may simply continue to displace *and* exacerbate environmental problems across time and space, further obscuring the complex, inequitable and evolving relationships that underpin current ecological and social dilemmas. The proposed 'tweaked contracts' also raise troubling ethical questions. For example, just because it is *possible* to trade carbon emission rights does not mean it is *desirable* to do so, especially if the establishment of such rights permits polluters to continue to emit greenhouse gases or use energy in manner that absolves them of responsibility for reducing energy consumption (Eckersley, 1993; Sandel, 2009).

Although alternative contractual responses to climate change have been discussed, they have seldom been attributed the attention or legitimacy of market-based solutions. For example, small Pacific Island states have called for a social compact that sets out the rights and obligations of parties in the context of historical injustices, the need for compensation transfers of wealth, technology and information, mitigation of future climate change and self-determination in adaptation decision making (Hayward, 2008; Lefale, 2008). This social compact asks the parties to secure a broader vision of human security than the narrow aspirations of the climate contracts described above, which simply tweak the objectives of the current Kyoto agreement. Commentator Andrew Sullivan (2009) also notes a growing backlash against a new top-down global climate contract in favour of bottom-up, context-sensitive local regulations and agreements. Varied climate justice groups advocate revisiting social justice issues in the principles of the United Nations Framework Convention on Climate Change (UNFCCC) by emphasising principles of common but differentiated responsibilities, intergenerational equity and polluter pays (Duxfield, 2007).

Such alternatives are difficult to realise when economic growth and economic prosperity are considered revered tenets of social contracts. Furthermore, it is not clear that new contracts are likely to provide sufficient checks on diffuse power relationships. For example, Dryzek (1987) points out that it is difficult for the state to develop policy that undermines the process of capital accumulation or for the state to subordinate the goal of economic growth to environmental objectives. Contesting social contracts in the name of the environment would require states to revisit the very processes of economic investment that have contributed to the exploitation of resources and disregard for ecosystem services.

Faced with new shocks, surprises and ongoing slow but non-linear change, traditional political institutions tend to respond by adopting known techniques, in this case more market tools and more strategies aimed at the creation and protection of wealth (Weber, 2008). Such approaches entirely disregard the need to consider and respond to the much deeper, systemic problems that underlie climate change. In other words, the 'tweaked' contracts that are currently being developed in the context of globalised economic power relationships are unlikely to secure equity or justice for marginalised groups or non-human nature. Given the privileged place that the goals of capital accumulation and economic growth have in social contract theory, there is a real danger that any new environmental contracts or 'new deals' will simply reinforce the power structures and economic relations that have contributed to an increase in greenhouse gas emissions, as well as rendered many individuals and communities vulnerable to the consequences of climate change. The nature of the deep injustices embedded in many social contracts and the complexity of climate change suggests that redefining the social contract is not a process that will occur inevitably, gracefully or spontaneously. Meaningful change will require contracts that facilitate deep structural transformations, and such responses seldom happen by accident or 'autonomously'. Consequently, a radical questioning of responses will most likely come from outside of the state and will require debate, discussion, struggle or conflict (Mouffe, 2005).

Alternative visions to market-liberal contracts

Despite the problematic legacy of market-liberal contract thinking, the rhetoric of social contracts does offer a powerful transformative inspiration for those seeking a more just future. Social contracts also provide an opportunity to articulate obligations and rights, which can be usefully employed to hold reluctant parties to account for their actions. Importantly, there are alternative traditions in social contract theory that can inform something new, such as multi-level and transformational compacts. The writings of eighteenth-century philosopher Jean-Jacques Rousseau, for example, can be reinterpreted in a contemporary context to inform responses to climate change. Likewise, a rich variety of alternative visions for managing human relationships exist, such as capability approaches (Delamonica and Mehrotra, 2006; Nussbaum, 2006); ecological citizenship (Dobson, 2003) and greened states (Eckerlsey, 2004), to name but a few. The capability approach developed by Amartya Sen and Martha Nussbaum, for example, can potentially serve as a basis for just, equitable and sustainable responses to climate change. Below, we consider how both Rousseau's ideas on social contracts and Sen and Nussbaum's capability approach may contribute to models of new, multi-level ecosocial compacts to promote human security in a changing climate.

Rousseau's thinking on The *Social Contract*, published in 1762, offers inspiration for a different type of climate contract, one which aims to secure and foster the capacity for communities and individuals to debate their own agreements on ways that their communities might flourish within ecological limits. While much of Rousseau's argument suffers from the constraints of his time (including misogyny, racism and environmental determinism) there is a quality of struggle within his work that resonates with today's search for better visions of what it means to secure a good life in the context of climate change. Rousseau wrote from the margins of a society that was, as today, fractured by deep inequality and unsustainable growth and militarism, at a time when it appeared that 'civil energies' were worn out. Although Rousseau's writings are incomplete as a guide to social contract thinking for large modern democracies, the inclusive, deliberative and liberating perspective that informs Rousseau's work provides kernels of alternative thinking. Below, we briefly highlight his insights on: (1) human flourishing and citizenship within ecological limits; (2) a more inclusive concept of 'we'; (3) enhanced public accountability across time and space; and (4) the notion of reflexivity. We then consider how these insights may combine with a capability approach to inform the development of eco-social compacts.

Human flourishing and citizenship within ecological limits

One of the hallmarks of modernity is development of a rational worldview that separates humans from the environment, promoting the idea of human control over nature and an unlimited potential for growth (Castree, 2005). A liberal political philosophy contributed to the notion that the market economy can successfully deal with scarcity through price mechanisms, and that environmental problems can be resolved by ecological modernisation and continued growth. However, Rousseau's vision of a social contract reminds us that human flourishing needs to occur within ecological limits, and that humanity is more than the mere sum of resource appropriation and utility seeking behaviours. Rousseau argued that humanity flourishes when given an opportunity to experience nature. His writings in *Emile* (1966) noted that the development of human imagination and spirit required direct exposure to nature and that humanity should show compassion to animals as sentient beings. He argued that the goodness of humanity is best fostered within agreed limits, when we choose 'to limit our desires ... when we can do more than we want, we are really strong' (Rousseau, 1968: 128; see also Rousseau, 1973).

An enhanced view of the 'we'

Climate change is a problem that requires recognition of the rights of, and responsibilities to, distant people and future generations, including vulnerable groups who have little voice in the social contracts of high consumption fossil fuel-based economies and societies (O'Brien *et al.*, 2009). This raises new questions of responsibility and compensation for citizens and governments that are not formal parties to a particular social contract within a clearly defined national boundary (Mueller, 2003). Although Rousseau's writing on the social contract was inevitably shaped by his time and the then-dominant paradigm, he nonetheless held a richer view of who should constitute the people, or parties to the social contract. Rousseau argued for collective responses, envisioning citizens entering into a social contract together to create a 'general will' which defends the person and goods of each member with the collective force of all (Rousseau, 1968: 60). This vision opens up the possibility of a new form of civil society in which the justice or morality of actions are debated amongst citizens and in which compassion can be extended to non-human nature and the rights and needs of future generations. Although Rousseau's call for participation in common deliberation based on a general will has been critiqued for its potential tyrannical implications, Iris Young (1989) and Carole Pateman (1988) have long argued that there is no need to conflate agreement to act in concert with a consensus and the tyranny of the majority.

Increasing public accountability across space and time

The global nature of climate change calls for alternative, decentred ways of making decisions that can be accountable to others across space and time. In his own work, Rousseau concluded that it was not possible to have accountable contracts beyond small local communities and city states. However, he usefully proposed a mechanism for enhancing the accountability of local contracts using tribunals to review decisions (Rousseau, 1968: 168). Rousseau's tribunal was not made of experts, but rather of lay community members who would reflect on issues. The concept opens the possibility of local, national and international lay bodies that are accountable to the communities that elect them to review the justice of decision making in practice, considering, for example, the rights of non-human nature and obligations to past and future generations. In light of experiences with international courts of human justice, local environmental review panels, indigenous rights tribunals or truth and justice commissions, citizen climate justice tribunals could provide a way of holding decision makers accountable across space and time (Sharp, 1990; Rotberg and Thompson, 2000). Rousseau's concept of the tribunal is suggestive of possibilities for inter-linked deliberative discussions about climate justice that could be conducted from local to international levels. Such an approach may resonate with the demands of non-governmental groups for more procedural justice and inclusive deliberation, inspired by a richer perspective of human security than protection of the rights of the citizen as a 'consumer or appropriator' (Macpherson, 1973).

Reflexivity

Climate change is increasingly seen as an urgent issue that demands a rapid and comprehensive response. However, responses must be carefully assessed, for their consequences may create new vulnerabilities or contribute to new environmental, social, economic or political problems. There is a need for reflexivity in responding to global challenges. Rousseau argued that collective action was a necessity when humanity reaches a point where 'the human race will perish if it does not change its mode of existence' (Rousseau, 1968: 59). Yet, he cautions that collaborative agreements are unlikely to be just or effective if they are agreed to in times of fear. He warns that 'usurpers always choose troubled times to enact, in the atmosphere of general panic, laws which the public would never adopt when passions were cool' and thus argued that covenants, compacts or contracts should be entered into after calm reflection (Rousseau, 1968: 95). Pelling and Dill (2006) similarly observe that while disasters may open opportunities for a review of the adequacy of existing political arrangements, the social learning that occurs in the wake of these disasters rarely results in inclusive, just decision-making procedures or more rights for previously marginalised groups. In terms of human security, the responses to climate change may matter as much as the direct impacts.

Rousseau's insights provide a basis for rethinking social contracts in the face of a changing climate, and in particular they suggest moving beyond market-liberal contracts that perpetuate many of the processes that contribute to human insecurity. C. B. Macpherson (1973: 4) notes that Rousseau's work was part of a much longer legacy within social contract theory, informed by a vision of humanity where man [sic] is not seen as a consumer, but rather:

as a creator and enjoyer of human attributes, and that these include: the capacity for rational understanding, for moral judgment and for action, for ascetic creation or contemplation, for emotional activities of friendship and love, and, sometimes, for religious experience . . . man is not a bundle of appetites seeking satisfaction but a bundle of conscious energies seeking to be exerted whether that Western tradition is traced back to Plato or Aristotle or to Christian natural law, it is based on the proposition that the end or purpose of man is to use and develop his uniquely human attributes or capacities.

A somewhat similar vision of humanity is captured in the capability approach developed by Amartya Sen and Martha Nussbaum (Nussbaum and Sen, 1993; Sen, 1999, 2009; Nussbaum, 2006), which can be considered an alternative to the social contract tradition. Nussbaum and Sen are critical of the contractarian tradition of which Rousseau is a part, although we argue that his vision of the social and environmental nature of the contract is wider than that of many other contract theorists. Sen and Nussbaum, however, prefer the idea of social development as providing the opportunity to foster human capabilities, i.e. what people are actually able to do and be. They suggest human capabilities as a normative framework for evaluation of social arrangements, and for the design and assessment of policies and proposals about social change in society (Robeyns, 2003). The capability approach takes a broader view of humanity, including both objective and subjective dimensions, which have often been ignored in development policies and approaches informed by traditional welfare economics. The capability approach avoids an over-emphasis on security as materialism, and focuses instead on self-defined well-being, development as freedom and the extent to which actions improve the outcomes of any particular situation of injustice (Nussbaum and Sen, 1993; Sen, 1999, 2009; Nussbaum, 2006). As described by Robeyns (2003: 7–8), within the capability approach, '[d]evelopment and well-being are regarded in a comprehensive and integrated manner, and much attention is paid to the links between material, mental, spiritual and social well-being, or to the economic, social, political and cultural dimensions of life.'

The normative, embedded framework of a capabilities approach, combined with the insights on transformative social contracts from the writings of Rousseau, presents alternatives to market-liberal social contracts as solutions to climate change. These alternatives also remind us that all voices need to be heard in a process of imagining new social arrangements that might enable deep structural transformations and enhance the human security outcomes of particular communities in the face of climate change.

Towards a transformative eco-social compact

Social contract theory provides a powerful, if problematic ideal for organising collective action. Many tweaked market contracts forged within this tradition appear to promise certainty, and to clarify new roles, shared responsibilities and mutual obligations in an era of rapid change and dangerous environmental risk. However, much of social contract theory has also been complicit in creating and compounding ecological crises, of which climate change is but one increasingly visible manifestation. In a globalised, densely interconnected world, even quite small and seemingly defined groups of stakeholders can have profound and unanticipated effects on distant others. In a changing world, new ecological contracts need to meet pragmatic as well as moral and ethical tests. They must be flexible enough to meet new demands including, for example, protection of future generations who may not have been recognised in original agreements. They must also be able to be reviewed and amended in the light of new information (for example, increased risk of rapid sea level rise or rapid release of greenhouse gases from permafrost). Finally, they must be robust enough to ensure that both the benefits and burdens of change processes are fairly distributed among all parties. To be relevant in times of turbulence and uncertainty, however, any new contracts must also be based on agreed principles of procedural justice and inclusive debate about their embedded moral values – rather than emphasising only distributional justice and protection from environmental risk. In short, the nature of contemporary global problems requires us to re-conceive both the state and the citizen in more imaginative, ethical ways *and* to consider ways citizens and states might debate the questions of justice embedded in climate change.

Recent writing by green political theorists Andy Dobson and Robyn Eckersley suggest ways in which we might begin develop alternative arrangements that protect human capacities to flourish, and enable us to debate the injustices of a changing climate. Dobson (2003) explicitly rejects the market contract idea of citizens entering into reciprocal bargains (e.g. 'I do this for you, in the expectation that you will also act'). Instead, he calls for citizens in countries with unsustainable levels of consumption to act to reduce their level of consumption 'asymmetrically', because they recognise their choices have potential to affect the life chances of distant and unknown others (Dobson, 2003: 47). Eckersley's work similarly helps us to rethink the role of the state in a greened social contract. Eckersley argues that constitutional agreements should take into account ways in which the state can foster and promote discussion about humans flourishing *within* agreed ecological limits. In this process she envisages states becoming agents promoting trusteeship for a common good (Eckersley, 2004: 50).

Dobson and Eckersley's work suggest some of the many ways that we might begin to develop an alternative to market contracts, which we describe here as eco-social compacts. These compacts entertain the possibility for citizens to re-envisage their relationships to each other and the natural world, developing their own agreements and pathways for development. These compacts are not centred, but instead exist within cross-cutting networks of public debate and authoritative review to ensure that local agreements do not become exclusionary, irrelevant or unjust contracts. They represent stronger, inclusive, more participatory and deliberative arrangements that are better able to enhance the resilience of the socioeconomic systems by providing citizens with meaningful and effective opportunity to contribute to collaborative problem solving within ecological limits. We envisage eco-social compacts as agreements reached through decentred public discussions, at multiple levels, about what should be secured, for whom and how (Hayward, 2008).

In conclusion, we eschew a single grand 'climate contract' approach to solving the complex problems associated with anthropogenic climate change, or elegant international carbon market trading regimes, in favour of humbler, tentative, messy, interlinked public debates about ways to reach local climate change agreements that address injustices. Eco-social compacts are forged amongst citizens, NGOs, and between states through open public debate. These multi-level compacts for governance must be matched with tribunals and fora that broaden opportunities for wider public scrutiny to ensure such agreements are not rendered irrelevant, or create new injustices. Although such decentred eco-compacts can be useful policy tools, they should be applied cautiously, and reflexively, mindful of varying cultural contexts, and uncertain yet significant environmental outcomes and risks (Freeden, 2009).

A single climate treaty, local compact or regional agreement alone will not transform the development pathways that society is locked into. A more open and inclusive public dialogue, followed by actions that continually contest existing power relationships, may however create the foundation for promoting greater human security in a warming world. In a decentred world of complex, multi-level power relationships, eco-social compacts are useful and suggestive of new relationships, yet they will not be enough. Effective compacts also require leadership at all levels, joined-up regulatory environments, cross-cutting opportunities for pubic debate and appeal, and significant social learning with ongoing citizen scrutiny. We argue that this meshed solution is more likely to produce the deep, systemic, complex behaviour and value transformations required to address climate change.

Acknowledgements

We would like to thank Fikret Berkes for inspiring our thinking on social contracts in relation to the resilience of social-ecological systems, and we are grateful to Michael Freeden and Andy Dobson for useful comments on earlier drafts.

References

Adger, W. N. and Jordan, A. 2009. *Governing Sustainability*. Cambridge, UK: Cambridge University Press.

Adger, W. N., Paavola, J., Huq, S. and Mace, M. J., eds. 2006. *Fairness in Adaptation to Climate Change*. Cambridge, MA: MIT Press,

Barry, B. 1995. *Justice as Impartiality*. Oxford, UK: Clarendon Press.

Bohman, J. 2004. Decentering democracy: inclusion and transformation in complex societies. *The Good Society*, **13**(2), 49–55.

Boucher, D. and Kelly, P. 1994. The social contract and its critics. In D. Boucher and P. Kelly, eds., *The Social Contract from Hobbes to Rawls*. London: Routledge, pp. 1–34.

Cash, D. W., Adger, W. N., Berkes, F. *et al.* 2006. Scale and cross-scale dynamics: governance and information in a multilevel world. *Ecology and Society*, **11**(2), 8. Available online: http://www.ecologyandsociety.org/vol11/iss2/art8/.

Castree, N. 2005. *Nature*. London: Routledge.

Cragg, W. 2000. Human rights and business ethics: fashioning a new social contract. *Journal of Business Ethics*, **27**(1–2), 205–14.

De Boer, Y. 2009. *Energy, Development and Climate Change. Address to the World Bank Group Energy Week*. Washington, DC, 31 March – 2 April, 2009. Available online: http://unfccc.int/files/press/news_room/statements/application/pdf/090331_speech_energy_wb.pdf.

Delamonica, E. and Mehrotra, S. 2006. *A Capability Centred Approach to Environmental Sustainability*. Monograph. Brazil: UNDP and International Poverty Centre.

Dobson, A. 2003. *Citizenship and the Environment*. Oxford, UK: Oxford University Press.

Dobson, A. and Eckersley, R. (eds). 2006. *Political Theory and the Ecological Challenge*. Cambridge, UK: Cambridge University Press.

Dryzek, J. 1987. *Rational Ecology: Environment and Political Economy*. New York, NY: Basil Blackwell.

Duxfield, F. 2007. The twin crises of climate change. *Just Change, Critical Thinking on Global Issues: Special issue Going Under*, **10** (October), 6–10.

Eckerlsey, R. 1993. Free market environmentalism: friend or foe? *Environmental Politics*, **2**, 1–20.

Eckersley, R. 2004. *The Green State: Rethinking Democracy and Sovereignty*. Cambridge MA: MIT Press.

Edenhofer, O., Luderer, G., Flachsland, C. and Fussel, H. 2008. *A Global Contract for Climate Change*. Background Paper for A Global Contract for Climate Justice Conference of the European Parliament Brussels, 11 November 2008. Available online: http://global-contract.eu/content/file/GlobalContract_Backgroundpaper.pdf.

Freeden, M. 2009. Failures of political thinking. *Political Studies*, **57**, 141–64.

Gauthier, D. 1986. *Morals by Agreement*. Oxford, UK: Oxford University Press.

Green New Deal Group 2009. *A Green New Deal: Joined-up Policies to Solve the Triple Crunch of the Credit Crisis, Climate Change and High Oil Prices*. First Report of the Green New Deal Group. London: New Economics Foundation. Available online: http://www.neweconomics.org/gen/uploads/ 2ajogu45c1id4w55tofmpy5520072008172656.pdf.

Gundimeda, H. 2004. How 'sustainable' is the 'sustainable development objective' of CDM in developing countries like India? *Forest Policy and Economics*, **6**(3–4), 329–43.

Harsanyi, J. C. 1976. *Essays on Ethics, Social Behavior and Scientific Explanation*. Dordrecht, Holland: Springer.

Hayward, B. 2008. Let's talk about the weather: decentering democratic debate about climate change. *Hypatia*, **23**(3), 79–98.

Hobbes, T. 1998 (1651). *The Leviathan*. New York, NY: Prometheus Books.

Hurrell, A. 2006. The state. In A. Dobson and R. Eckersley, eds., *Political Theory and the Ecological Challenge*. Cambridge, UK: Cambridge University Press.

Jackson, T. 2009. *Prosperity Without Growth? Economics for a Finite Planet*. London: Earthscan.

Kissi-Mensah, L. 2008. The Impact of the Minerals Industry on Surrounding Communities, Specifically the Socio-Economic and Environmental Impacts, Factors and Responsibilities. *AUSIMM New Leaders Conference Australasian Institute Of Mining and Metallurgy 2008*, **6**, 23–5.

Kant, I. 1959 (1785). *Foundation of the Metaphysics of Morals*. New York, NY: Bobbs-Merrill.

Latta, A. and Garside, N. 2005. perspectives on ecological citizenship. *Environments*, **33**(5), 1–9.

Lefale, P. 2008. Beyond Science: Climate change as a perfect political dilemma. In B. Hayward, eds., *The Politics of Climate Change: Issues for New Zealand and Small States of the Pacific*. Auckland and Wellington, New Zealand: Dunmore Books.

Locke, J. 1965. *Two Treaties of Civil Government*. New York, NY: New American Library.

Macpherson, C. 1973. *Democratic Theory: Essays in Retrieval*. Oxford, UK: Clarendon Press.

Midgley, D., ed. 2005. *The Essential Mary Midgley*. London: Routledge.

Midgley, M. 2003. *The Myths We Live By*. London: Routledge

Miliband, D. 2006. *The Great Stink: Towards an Environmental Contract*. Audit Commission Annual Lecture, 19 July, London. Available online: http://www.eeph.org. uk/energy/index.cfm?mode=view&news_id=641.

Mills, C. W. 1997. *The Racial Contract*. Ithaca, NY: Cornell University Press.

Mouffe, C. 2005. *On the Political: Thinking in Action*. London/New York: Routledge.

Mueller, D. 2003. *Public Choice III*. New York, NY: Cambridge University Press.

Nussbaum, M. C. 2006. *Frontiers of Justice: Disability, Nationality, Species Membership*. Cambridge, MA: Harvard University Press.

Nussbaum, M. and Sen, A. 1993. *The Quality of Life*. Oxford, UK: Oxford University Press.

O'Brien, K. L., Hayward, B. M. and Berkes, F. 2009. Rethinking Social Contracts: Building Resilience in a Changing Climate *Ecology and Society*, **14**(2), 12.

Pateman, C. 1988. *The Sexual Contract*. Stanford, CA: Stanford University Press.

Pateman, C. and Mills, C. 2007. *Contract and Domination*. Cambridge, UK: Polity Press.

Pelling, M. and Dill, K. 2009. Disaster politics: Tipping points for change in the adaptation of socio-political regimes. Forthcoming in *Progress in Human Geography*, **34**(1), 21–37.

Pelling, M. and Dill, K. 2006. Natural disasters as catalysts for political action. *Chatham House ISP/NSC Briefing Paper*, **06**/01, 4–6.

Rawls, J. 1971. *A Theory of Justice*. Oxford, UK: Oxford University Press.

Robeyns, I. 2003. The capability approach: an interdisciplinary introduction. Available online: www.capabilityapproach.com/pubs/323CAtraining20031209.pdf.

Rotberg, R. and Thompson, D. 2000. *Truth Versus Justice*. Princeton: Princeton University Press.

Rousseau, J. 1973. (1762) *The Social Contract and Discourses*. Translated by G. Cole, J. Brunfitt and J. Hall. London, UK: Everyman.

Rousseau, J. 1968. (1762) *The Social Contract*. Translated by Maurice Cranston. Harmondsworth, UK: Penguin Books.

Rousseau, J. 1966. (1762) *Emile*. Translated by B. Foxley. London: Everyman.

Sandel, M. 2009. *Markets and Morals*. The Reith Lecture Series. London: BBC. Available online: http://www.bbc.co.uk/programmes/b00l0y01.

Scanlon, T. 1998. *What we Owe to Each Other*. Cambridge, MA: Harvard University Press.

Sen, A 1999. *Development as Freedom*. New York: Knopf.

Sen, A 2009. *The Idea of Justice*. London: Allen lane.

Sharp, A. 1990. *Justice and the Maori*. Auckland: Oxford University Press.

Sippel, A. 2009. Back to the Future: Today and Tomorrow's Politics of Degrowth Economics (Decroissance) in Light of the Debate over Luxury among Eighteenth and Early Nineteenth Century Utopists. *International Labour and Working Class History* **75** (Spring): 13–29.

Simms, A. 2008. New Green Deal for Our Times. *The Observer*, 3 August 2008. Available online: http://www.guardian.co.uk/business/2008/aug/03/economicgrowth. climatechange.

Sullivan, A. 2009. It's the little things that will fight climate change. *The Times*, 5 April 2009. Available online: http://www.timesonline.co.uk/tol/comment/columnists/andrew_sullivan/article6035799.ece

Walzer, M. 1998 The Civil Society Argument. In G. Shafir, ed., *The Citizenship Debates*. Minneapolis, MN: The University of Minnesota Press, pp. 291–308.

Weale, A. 2004. Contractarian theory, deliberative democracy and general agreement. In K. Dowding, R. Goodin, C. Pateman and B. Barry, eds., *Justice and Democracy: Essays for Brian Barry*. Cambridge, UK: Cambridge University Press, pp. 79–96.

Weber, E. 2008. Facing and managing climate change: assumptions, science and governance responses. *Political Science*, **60**(10), 133–49.

White, A. L. 2007. Is it time to rewrite the social contract? *Business for Social Responsibility*. Available online: www.tellus.org/publications/files/BSR_AW_Social-Contract.pdf.

Young, I. 1989. Polity and group difference. *Ethics*, **99**(2), 250–74.

Young, I. 2006. De-centering deliverative democracy. *Kettering Review*, **24**(3), 43–60.

Zadek, S. 2006. *The Logic of Collaborative Governance: Corporate Responsibility, Accountability and The Social Contract*. Paper 17, The Corporate Responsibility Initiative, John F. Kennedy School of Government, Harvard University, Cambridge, MA.

12

Towards a new science on climate change

KAREN O'BRIEN, ASUNCIÓN LERA ST.CLAIR AND
BERIT KRISTOFFERSEN

Introduction

Over the past two decades, human security has developed into an important international discourse that draws attention to the well-being of individuals and communities in the face of multiple stressors and threats. By embracing both normative
and ethical perspectives, human security draws attention to the factors that influence
the capacity of individuals and communities to respond to threats to their needs,
rights and values (Barnett *et al.*, 2010). Drawing on moral and philosophical
arguments, this book shows how human security can serve as a critical lens through
which climate change can be discussed, analysed and addressed. It allows us to
inquire, assess and evaluate climate change processes and their outcomes from the
perspective of what matters to human beings, both individually and collectively.
From such a perspective it is possible to link environmental changes directly to the
factors that create and perpetuate poverty, vulnerability and insecurity. Issues of
power, politics and interests inevitably arise, but so do questions of culture, values,
beliefs and worldviews. Human security emphasises not only how humans individually and collectively experience climate change, but also how they perceive their
responsibilities towards future generations, including their own capacity to forge
outcomes that can build a more sustainable and equitable future.

The prevailing discourse on climate change frames it as a serious environmental
problem that will affect humanity in unprecedented ways if it is not immediately
addressed. There is an implicit understanding that everyone's security is at stake,
including the security of the human race; yet, there is also recognition of differential
vulnerability and adaptive capacity. People in the global South are already the most
affected by climate change; the most vulnerable in the future are likely to be the poor
people and poor regions of the planet (UNDP, 2007/2008). While this environmental discourse makes it clear that climate change will have uneven impacts, it
fails to critically examine the social, institutional, economic, political and human

Climate Change, Ethics and Human Security, eds. Karen O'Brien, Asunción Lera St.Clair and Berit Kristoffersen.
Published by Cambridge University Press. © Cambridge University Press 2010.

context in which anthropogenic climate change is created and experienced; and it fails to critically analyse the models of development and progress that have produced the climate crises in the first place. These fundamental failures have wide implications; they constrain responses to technological and managerial approaches that often address only the symptoms, and they may facilitate political games around climate negotiations that can limit public debates to assessments of costs and benefits, rather than promote the wider ethical debates that the issue deserves. In this concluding chapter, we therefore discuss how human security perspectives can contribute to a shift in the discourse, and the framing of a new science on climate change.

The institutionalised discourse on climate change

The IPCC Fourth Assessment Report (2007) represents a milestone in climate change history. Building on previous reports published in 1990, 1995 and 2001, the three volumes of the Fourth Assessment draw attention to both the certainty and severity of climate change and associated impacts, and to the necessity for both adaptation and mitigation as responses to climate change. The three working groups of the IPCC have assessed the science of climate change, reviewed impacts, vulnerability and adaptation, and considered current and potential policy responses towards reducing greenhouse gas emissions. The IPCC reports have also helped to situate droughts, floods, wildfires and some record-breaking extreme events within the context of long-term changes linked to increased atmospheric concentrations of greenhouse gases (IPCC, 2007). Research shows that many sectors and regions are vulnerable to climate change, and that present and future achievements linked to welfare and sustainable development are threatened by climate change (IPCC, 2007; UNDP, 2007/2008; UNISDR, 2009).

The IPCC, as a recipient of the 2007 Nobel Peace Prize, has successfully raised climate change as an issue of peace and security. It is, without a doubt, the knowledge body with sufficient legitimacy to assess the problem of climate change. Yet, the question of what to do about climate change is another issue. The experts that work on the IPCC reports cannot provide neutral and objective advice as to what kinds of trade-offs are fair; nor can they objectively interpret what climate change means to individuals and members of communities. Normative and ethical questions cannot be 'assessed' in the same way that, for example, research on sea level rise can be assessed. Questions as to which actions to prioritise, who should pay for the costs of adaptation, or how to compensate groups for the loss of livelihoods and culture, require deliberation and debate among members of society with different interests and prioritised values. Many of the issues surrounding climate change cannot be resolved by expert knowledge alone.

Scientists are now working on the framing of the IPCC's Fifth Assessment Report (IPCC, 2009). It is worth questioning, however, whether many of the issues raised in this book can be adequately addressed by the IPCC as it is organised today. Although the IPCC attracts excellent scholars, it is neither structured nor mandated to assess critical research from the social sciences, the humanities or disciplines that present critical perspectives on human security, development and poverty. The IPCC's mandate is to synthesise knowledge about *climate change* and not to open up for an assessment of the cross-cutting processes that contribute to climate change impacts and vulnerability, or facilitate or constrain responses.

It is important to point out that the world does not have a global institution akin to the IPCC to assess the state-of-the-art knowledge on poverty and development. As discussed in Chapter 8, the presumption that multilateral bodies can serve as counterparts to the IPCC on these issues is misguided. The World Bank is already seen by many as the 'default' knowledge institution for work related to poverty and development, and many expect it to play a dominant role in implementing any agreement on adaptation that is likely to emerge from the 2009 climate change negotiations in Copenhagen. While multilateral institutions certainly have a role to play in addressing climate change, the actual role needs to be openly debated, particularly since successful responses involve transforming the very ideas of progress and development that have led to climate change in the first place. Many of the key actors working on development and poverty reduction lack reflexivity regarding the dominant discourses and research framings. Indeed, one of the fundamental problems with much of the multilateral work on poverty and development is its narrow, economistic and highly politicised focus. An over-emphasis on politicised economic perspectives is likely to undermine rather than promote fair and open global debates. To prevent irreversible changes in the global climate system while reducing vulnerability to changes that are inevitable requires an integration of disciplines and perspectives. This includes normative perspectives that prioritise human security and questions of global justice for present and future generations.

A new science on climate change

What is needed is a 'new science' on climate change – one that is able to integrate insights from the social sciences, the humanities and other fields with emerging findings on how human activities influence the Earth System, and that can place issues of justice, ethics, responsibility and human security at the forefront of policy debates. The new science can involve anthropologists, psychologists, historians, linguists, media researchers, lawyers, philosophers and the many other experts in the fields that the IPCC reports have not thoroughly assessed. This new science should also listen to the voices of activism, emergent legislation and the multi-voiced political discourse at all scales. It must recognise that many of the issues raised by climate change cannot be addressed by scientific or expert responses, and that many of the challenges and unavoidable trade-offs are going to have to be debated in spaces of public deliberation.

There are many places from which to start building this new science, including from some of the issues raised by post-normal science (Funtowicz and Ravetz, 1993) and from the work on sustainability science (Kates *et al.*, 2001). There is already a strong foundation for assessing why climate change matters; and what is really at stake. Numerous questions that are currently being addressed by research-ers in the social sciences and humanities draw attention to both subjective and objective dimensions of global change processes, which together can be used to assess the 'so what?' of climate change. For example, how does climate change influence the capacity of individuals and communities to respond to multiple and interacting stressors? Whose security is most threatened by climate change and why? What responsibilities do diverse actors have and how can they be activated? What types of adaptation are sustainable, and what types contribute to the vulner-ability of others and of future generations? How does climate change influence the diversity of needs and values that contribute to human security? What are the cultural implications of climate change and how does this influence world views and belief systems? Whose values count in contemporary responses (or lack of responses) to climate change and how can value conflicts be resolved?

A wider normative basis for understanding and addressing climate change will be a key element of the new science. The new research agendas we propose here point out that climate change is more than an environmental problem – it is a social problem, a development problem and an ethical problem that is closely linked to the security of humankind. Thus climate change impacts do not result from changing climate parameters alone, but instead they are intensified, reduced or eliminated by the context within which the changes take place. A new science on climate change must make it clear that responses to climate change must extend beyond 'climate policies' to address the social, economic, institutional and political context in which climate change is occurring and experienced. Some of the key themes that might be included are considered below.

The social context

There is sufficient knowledge to conclude with confidence that those who would suffer the most from climate change are those who are already experiencing the negative impacts of other global challenges, such as the financial/economic crisis, ongoing conflicts, environmental degradation and loss of biodiversity (Leichenko and O'Brien, 2008; Dalby, 2009). They also include people with inadequate housing, those who lack access to health, education, clean water and those living under conditions of food and labour insecurity. Significantly, those who are least likely to be able to respond to future changes are the ones who have contributed least to the problem in the first place, and who have benefited least from modernisation and industrialisation. The social context of climate change can be better understood by assessing critical poverty studies, rather than the dominant perspectives' proposed by governments and multilateral institutions (Lawson and St.Clair, 2009).

As many of the chapters in this book have documented, climate change is one among many processes that create shocks and stresses for individuals and communities. Both the climate change and disaster risk reduction communities are aware that impacts and vulnerability are determined not by the magnitude of the shock or stressor alone, but by interactions with other processes, particularly those that influence livelihoods and development (Schipper and Pelling, 2006). For example, three years of consecutive drought may be much more significant for individuals or communities whose livelihoods are threatened by import competition than for individuals who have diversified their livelihood options. In some cases, a small change or a single extreme event may push people into a situation of insecurity and deteriorating well-being, while in other cases, individuals may be able to maintain security in the face of rapid or major change.

What is missing from many assessments of climate change is a critical review of how and why the social context is changing as a result of interactions among ongoing and emerging processes, often creating the context for disasters and major humanitarian crises. A new science would problematise the systematic factors that make many people vulnerable to change. Negative outcomes associated with climate change in turn affect the social context and can influence processes such as peace-building, democratisation and fair and equitable development. This new science would assess when, how and why society is becoming less resilient to change over time, and identify ways to increase resilience in sustainable and equitable ways. Understanding the social context in which climate change is both created and experienced involves going beyond the development of socioeconomic scenarios for the future and towards a critical analysis of the social, economic and political structures that define the context.

The institutional context

Institutions and institutional capacity are considered prerequisites for responding to climate change. For example, the IPCC Fourth Assessment Report (2007) highlights their importance as a means of mitigating climate change, reducing vulnerability and increasing adaptive capacity. The institutional constraints to adaptation were discussed, including the difficulties in mainstreaming adaptation into development planning (Adger *et al.*, 2007). Economic development, changes to ecosystem services, urbanisation, migration, macro-economic policies and many transformations linked to globalisation processes were recognised to influence the institutional context for responding to climate change and extreme events.

The financial crisis has illustrated many of these institutional dynamics, and it reveals how complex institutional arrangements have contributed to a loss of resilience within and across diverse social–ecological systems (Leichenko *et al.*, 2010). Indeed, recent newspaper headlines have drawn attention to the unravelling of a large and complex financial system and its implications for social welfare and human well-being through impacts on production, jobs, investments, development aid and humanitarian assistance. This not only contributes to uncertainty about the future, but it also affects climate change adaptation and mitigation efforts, as well as disaster risk reduction strategies.

What is missing from the current debates about climate change is an assessment of the dynamic institutional context in both developed and developing countries, including new linkages (e.g. financial, technological, informational) that are connecting vulnerability across distant places and groups, and in many cases transferring vulnerability to future generations. A new science on climate change would address the changing institutional contexts for dealing with risk, vulnerability, uncertainty, fairness and compensation issues. For example, it might reassess the role of social contracts in a changing climate, including how the notion of rights and responsibilities between states and citizens is changing as risks become increasingly global (see Hayward and O'Brien, Chapter 11). It would consider to what extent social institutions, like the family, which have been dramatically changed by post-industrial transformations and globalisation processes, would be affected by climate change-related stresses. The new science would go deeper into the gender perspectives of climate change, and into how social exclusions rooted in existing institutions are reinforced by uncertainty and change. Moreover, it would explore the extent to which institutions at various scales are actually able to understand climate science and to integrate that knowledge into adaptive responses. The new science would assess both the dynamics and resilience of the institutional context and point to the types of changes that might facilitate responses to climate change that enhance rather than compromise human security.

The human context

Climate change research has drawn attention to the impacts of climate change on different regions, ecosystems and sectors, and considered the factors that contribute to vulnerability and influence adaptation. However, little attention has been given to the subjective dimensions of climate change, including how religion, spirituality, culture, values, world views and belief systems influence the perceived outcomes of climate change and responses to climate change. Climate change will affect not only human lives and material needs, but also experiences and relationships that may be valued differentially by individuals, communities and cultures, both in the present and the future (O'Brien, 2009). It will affect and challenge existing beliefs and world views and alter rituals, practices and the many different senses of 'belonging' to a particular place or region. The emotional consequences of climate change have been largely ignored within the dominant discourse, despite knowledge that climate-related disasters can have long-lasting development effects, as experienced, for example, through post-traumatic stress disorder. Understanding the 'interior' human context in which climate change is created and experienced is essential to any efforts to reduce vulnerability and mitigate the long-term consequences of climate change (O'Brien and Hochachka, 2010; O'Brien and Wolf, 2010).

What is missing from most climate change research, including from the rapidly emerging climate and development discourse, is a broader assessment of the implications of climate change for human security, i.e. a deeper understanding of what reduces or enhances the capacity of individuals and communities to respond to threats to their social, environmental and human rights. The new science that we argue for would draw attention to how climate change may affect the subjective dimensions of individuals and collective experiences, including relationships with species, ecosystems, cultural icons and so on, that people value (Adger *et al.*, 2009). Although many of the subjective and ethical dimensions of both development and climate change have been studied, these strands of research have not been integrated with understandings of climate change impacts and adaptation.

Climate change itself challenges some firmly entrenched belief systems and understandings of human–environment relationships and the future of the human race (O'Brien and Hochachka, 2010). The notion that humans can influence a complex climate system is in many ways a radical departure from traditional, hierarchical world views that attribute changes to external or supernatural forces, and even from modern world views that emphasise a dualism between humans and nature, along with the idea that nature can be effectively controlled through technology and management alone (Castree, 2005). Ideas about whether humans can influence the future climate by reducing greenhouse gas emissions, adapt to climate variability and change, and successfully address poverty and inequality, are as much about beliefs as they are about the science of global change.

An understanding of the subjective and ethical aspects of climate change is of fundamental importance to creating an alternative future. How do we forge a sense of equitable global agreements? How do we promote solidarity in responses and fight self-interest and short-sighted emotional reactions? How can we forge a culture of shared responsibility? These and many other related questions cannot be addressed by scientific and economistic views on climate change, but rather they must be understood as part of the larger human context.

Human security in a global perspective

The concept of human security draws attention to individuals and communities, rather than the state, as the focus of security concerns (Barnett *et al.*, 2010). Yet history shows that the security of some has often been realised at the expense of the security of others. In order to build a culture of shared responsibility, we must start with framing the issues in ways that build and forge rather than prevent a sense of unity across cultures, groups and interests. This calls for concepts and ideas that challenge the dominant Western ideological and methodological hegemony that pervades political process of consensus building and creates tensions in North–South relations. To realise human security at a global scale requires both conceptual as well as real spaces for sorting out conflicts of values, and for redefining interests, rights and responsibilities in terms of the largest possible interpretation of 'we'. Such upscaling is crucial to preventing the 'dumping' of negative climate change impacts onto distant others and onto future generations.

Human security, ethics and moral philosophy are currently dominated by Western traditions of thought. Yet much has been achieved in debates on other global issues, such as in the field of development ethics, that can contribute to more inclusive perspectives. Development ethics has, for example, produced much work exploring the different meanings, experiences and evaluations of development and under-development, including the ways in which models of progress that have disasso-ciated humans from nature tend to be less effective in leading towards equitable development (Gasper and St.Clair, 2010). Most importantly, one of the fundamental lessons learned from decades of work on value-based and rights-based approaches to poverty reduction and development is that reasoning and deliberation are indeed possible, even when the task seems overwhelming and unprecedented in scope and in intensity. We can also learn from earlier work on the difficulties and risks of introducing justice and ethical concerns into the work of major development agencies, and to tap into such knowledge for better influencing them in regards climate change (McNeill and St.Clair, 2009). The growing field of climate ethics suggests that human security and ethical perspectives can indeed serve as the basis for a new science for climate change (see http://climateethics.org). Human security and ethical perspectives on climate change can be used to create a multi-voiced arena for debate and deliberation, contributing to greater transparency in terms of whose voices are heard, whose views have more epistemic values, and which ones are the sources of power in dominant discourses on climate change.

Much work remains to be done to integrate the beliefs, values and aspirations of diverse individuals and communities into a coherent vision for the future – particularly if this vision is to include global-scale responses to climate change that prioritise human security. This calls for including and respecting diverse values, rather than arguing for one single dominant value perspective. Yet it also requires resolving value conflicts in an ethical and fair manner that takes into account new understandings of the relationships between humans and the environment that are emerging from Earth Systems science, philosophy, etc. The normative concept of human security provides some simple ethical guidelines for the types of responses that should be prioritised, i.e. those that enhance the capacity of individuals and communities to respond to threats to their environmental, social and human rights, both in present and future generations.

Climate change, ethics and human security

A key argument of this book is that the knowledge that is needed to inform discussions and actions in response to climate change needs to be broadened. Part of the challenge is to coordinate and synthesise research and insights from the many actors across the world who are concerned with these issues, and who are producing important and relevant knowledge, including ethical perspectives on climate change. To advocate a new way of thinking means seeing the connections between the causes and consequences of climate change and present and future vulnerabilities, insecurities and injustices. It points to new ideas about alternative futures, building on sources of knowledge that are often discounted or bypassed by standard scientific discourses and methodologies. A new science for climate change should be broader, more inclusive and address one fundamental question, namely, *how can we respond to climate change in a manner that enhances human security and promotes global justice in an unequal but increasingly interconnected world?*

Currently, mention of ethical issues, human security or human rights perspectives on climate change occurs at a very rhetorical level, characterised by 'more talk than action'. Yet there are instrumental reasons for framing climate change as an issue of human security, and for forging a new science for climate change. The negative impacts of climate change, particularly in relation to the long-term deprivation of basic needs and the right to food, water and shelter, will add to an already visible and worrisome sense of resentment and injustice by those who have already been impacted by an unfair global system. These negative feelings and emotions may lead to violence, particularly against women, or to outright genocide as witnessed in cases such as Rwanda. Shifting the discourse towards a broader understanding of the problems and solutions calls for integrated expertise and for building knowledge that would permit a fair and democratic process for responding to closely interlinked global challenges.

The current framing, which considers climate change as an environmental issue, draws heavily on scientific, environmental and technological know-how. This framing has led to a narrow set of responses to climate change: as a matter of managing resources in a sustainable way, developing 'green' technology and collecting sufficient data to plot future scenarios of climate change. Promoting green economic development, trading greenhouse gas emissions, advocating carbon capture and storage and geoengineering, protecting the forests in far away lands and so forth, all sound like promising actions in response to climate change. However, without a serious challenge to technocratic ideas of progress and development, a questioning of current consumption patterns, and a closer examination of existing inequalities in power and resources, these actions can be considered a 'road to nowhere'. Worse still, they may represent a dangerous detour that favours the adaptation of the rich, while sanctioning the disappearance of the poor and vulnerable.

Although the responsibility to act lies with everyone – politicians, business leaders, teachers, citizens groups, communities and individuals, the scientific community has an important role to play in mediating knowledge and fostering a deeper understanding of the scientific complexities and societal consequences associated with climate change. Researchers and educators from the natural sciences, social sciences, humanities and other areas have an important responsibility to act as intermediary agents, proposing alternative pathways and tools that can help society to realise a shift in politics and practices – and to realise that a failure to respond is a strong political act that has enormous ethical implications.

The current failures in national and global governance to act upon climate change reflect more than a lack of political will, but incapacity to envision and move towards the creation of a fair world where the benefits of modernisation and progress are shared and the costs fairly distributed. We need nothing less than a complete rethinking of what progress is, what development means and what a just and fair world that protects and promotes human security might look like. This is the joint work of politicians and experts, of citizens and people from all over the world. Many of the questions raised by climate change have no scientific answers; all answers invariably raise new questions, including questions of how we perceive of and relate to one another.

If we fail to understand this basic insight, important decisions about the future will be left to power games, and when it concerns power, the values and interests of global elites generally win the arguments. The general public is often not exposed to the ways in which power politics dominate climate negotiations at the national and international levels. The wider public has not been invited to take part of or engage in discussions of the moral and ethical dimensions of climate change – this has been the purview of academic debates and deliberations. Meanwhile, the wider public in Western societies (but also many elites in developing countries) have become accustomed to thinking that matters of poverty and development are the business of development and humanitarian aid organisations, and not a shared responsibility.

We argue that climate change discussions should reflect the undeniable fact that there are winners and losers of both globalisation and climate change (O'Brien and Leichenko, 2003), and that the values and interests of singular nation-states, elites and powerful industries and groups should be made transparent in climate change negotiation processes. The powerful within these negotiations keep the framing of the issue in the realm of the scientific, environmental or technological dimensions – approaches that do not challenge the uneven, unfair and unsustainable growth and consumption patterns that are the core links to the climate crisis, the poverty crisis, the food crisis, etc. As much as humanitarian and development aid is necessary (and a lot more is and will be needed to address emergency situations of hunger, floods or droughts), responsibility for responding to climate change should not be placed with aid institutions alone, particularly with institutions that are still perpetuating the very patterns of development that have led to the climate crises while locking millions of people in poverty, without consideration of human rights and the development of human capacities and aspirations.

This focus on ethics and human security is not merely an academic exercise in rethinking concepts and reframing complex problems; it points towards a much broader capacity to find solutions and principles to guide us on difficult questions. Different world views, beliefs and values interact with relations of power and interests to influence the direction that society moves and the outcomes that are experienced. The possibility to self-reflect and to search for and develop increased options and capacities to respond to global threats and challenges is paramount. The capacity for self-reflection is, after all, one of the true signs of progress and development. Can the global community 'prevent the repetition of the most ancient pattern in human history: that real change never actually comes until after a crisis? Can the human community break out of this chain of cause and effect?' (Garrison, 2005: 207).

In short, framing climate change as an issue of ethics and human security opens up new debates, discussions, research questions, policies and possibly alliances that can contribute to resolving one of the biggest and most complex problems of the century. Society is challenged to transform itself and to recognise that human security is about more than individual or community well-being. It is also about a collective and connected state of well-being that is continually negotiated by and for individuals and communities who recognise that their actions, including responses to climate change, are about individual and collective responsibilities that can and will influence trajectories for human development. There is no doubt that the key challenges to increasing human security in a changing climate can be met. There are signs that the discourse is already shifting, and that the links between climate change, ethics and human security are becoming both visible and better understood. A new science on climate change can hasten this shift.

References

Adger, W. N., Agrawala, S., Mirza, M. M. Q. *et al.* 2007. Assessment of adaptation practices, options, constraints and capacity. In M. L. Parry, O. F. Canziani, J. P. Palutikof, P. J. Van der Linden and C. E. Hanson, eds., *Climate Change 2007: Impacts, Adaptation and Vulnerability. Contribution of Working Group II to the Fourth Assessment Report of the Intergovernmental Panel on Climate Change*. Cambridge, UK: Cambridge University Press, pp. 717–43.

Adger, N., Barnett, J. and Ellemor, H. 2009. Unique and valued places at risk. In A. Rosencranz, and S. Schneider, eds., *Climate Change Science and Policy*. Washington, DC: Island Press.

Barnett, J., Matthew, R. and O'Brien, K. 2010. Introduction: global environmental change and human security. In R. Matthew, J. Barnett, B. Macdonald and K. L. O'Brien, eds., *Global Environmental Change and Human Security*. Cambridge, MA: MIT Press, pp. 3–32.

Castree, N. 2005. *Nature*. London: Routledge.

Dalby, S. 2009. *Security and Environmental Change*. Cambridge, UK: Polity Press.

Funtowicz, S. and Ravetz, J. 1993 Science for the post-normal age, *Futures*, 25/7 September 1993, 735–55.

Garrison, J. 2005. America as empire: global leader or rogue power? In F. Dodds and T. Pippard, eds., *Human and Environmental Security: An Agenda for Change*. London: Earthscan, pp. 199–208.

Gasper, D. and St.Clair, A. L. 2010. Development ethics an emergent field: introduction. In D. Gasper and A. L. St.Clair, eds., *Development Ethics: A Reader*. London: Ashgate.

IPCC 2007: *Climate Change 2007: Impacts, Adaptation and Vulnerability. Contribution of Working Group II to the Fourth Assessment Report of the Intergovernmental Panel on Climate Change*, M. L. Parry, O. F. Canziani, J. P. Palutikof, P. J. van der Linden and C. E. Hanson, eds., Cambridge, UK: Cambridge University Press, 976 pp.

IPCC 2009. *Chairman's Vision Paper*. Submitted by IPCC Chairman to Scoping Meeting for the IPCC Fifth Assessment Report (AR5), Venice, Italy 13–17 July 2009. Available online: http://www.ipcc.ch/scoping_meeting_ar5/documents/doc02.pdf.

Kates, R., William, W., Clark, C. *et al.* 2001. Sustainability science. *Science*, **292**, 641–2.

Lawson, V. and St.Clair, A. L. 2009. *Poverty and Global Environmental Change*. International Human Dimensions Programme Update, Issue 2. Available online: http://www.ihdp.unu.edu/category/47?menu=61.

Leichenko, R. M. and O'Brien, K. L. 2008. *Environmental Change and Globalization: Double Exposures*. New York, NY: Oxford University Press.

Leichenko, R., O'Brien, K. and Solecki, W. D. 2010. Climate change and the global financial crisis: using the lens of double exposure to motivate new research. *Annals of the Association of American Geographers*, (forthcoming).

McNeill, D. and A. L. St.Clair. 2009. *Global Poverty, Ethics, and Human Rights: The Role of Multilateral Organisations*. London/New York: Routledge.

O'Brien, K. 2009. Do values subjectively define the limits to climate change adaptation? In W. N. Adger, I. Lorenzoni and K. O'Brien, eds., *Adapting to Climate Change: Thresholds, Values, Governance*. Cambridge, UK: Cambridge University Press, pp. 164–80.

O'Brien, K. and Hochachka, G. 2010. An integral approach to climate change adaptation. *Journal of Integral Theory and Practice*, **5**(1):89–102.

O'Brien, K. and Leichenko, R. 2003. Winners and losers in the context of global change. *Annals of the Association of American Geographers*, **93**(1), 99–113.

O'Brien, K. and Wolf, J. 2010. A values-based approach to vulnerability and adaptation to climate change. *Wiley Interdisciplinary Reviews: Climate Change*, (forthcoming).

Schipper, L. and Pelling, M. 2006. Disaster risk, climate change and international development: scope and challenges for integration, *Disasters*, **30**(1), 19–38.

UNDP (United Nations Development Programme) 2007/2008. *Fighting Climate Change: Human Solidarity in a Divided World*. New York, NY: United Nations Development Programme.

UNISDR 2009. *Global Assessment Report on Disaster Risk Reduction (2009)*. United Nations International Strategy for Disaster Risk Reduction (UNISDR). Geneva: UNISDR.

Index

228

Printed in the United States
by Bookmasters

Printed in the United States
By Bookmasters